A History of the Human Brain

A
HISTORY
of the
HUMAN
BRAIN

From the
Sea Sponge to CRISPR,
How Our Brain Evolved

BRET STETKA

TIMBER PRESS

Portland, Oregon

Published in 2021 by Timber Press, Inc.

The Haseltine Building
133 S.W. Second Avenue, Suite 450
Portland, Oregon 97204-3527
timberpress.com

Printed in the United States

Text design by Sarah Crumb
Jacket design by Adrianna Sutton and Jason Alejandro

ISBN 978-1-60469-988-3

Catalog records for this book are available from
the Library of Congress and the British Library.

For mom, dad, and Amanda

CONTENTS

Prologue 11

PART ONE | THE ANIMAL COLLECTIVE

1. Very Ape 17
2. Life from No Life 25
3. Fish, Head 39
4. Lizard Kings and Lemurs 55
5. Upright Citizens 69

PART TWO | OUR SOCIAL BRAIN

6. Groomsmen 99
7. A History of Violence 113
8. Our Softer Side 125
9. Speaker Wire 135
10. Temperament of the Dog 145

PART THREE | FOOD, FIRE, AND THE FUTURE OF THE HUMAN BRAIN

11. Weather Permitting 163
12. A Paleolithic Raw Bar 173
13. The Creatives 185
14. The New Wave 201
15. Future Sapiens 211

Acknowledgments 231
Notes 233
Bibliography 239
Photo and Illustration Credits 256
Index 257

The brain is a world consisting of a number of unexplored continents and great stretches of unknown territory.

—SANTIAGO RAMÓN Y CAJAL

I was taught that the human brain was the crowning glory of evolution so far, but I think it's a very poor scheme for survival.

—KURT VONNEGUT

PROLOGUE

Back in May of 2015 I spent a week in Canada reporting from the American Psychiatric Association's annual meeting.

While strolling the halls of the Metro Toronto Convention Centre at what I recall being one of those inhumanely early hours during which only doctors and shift workers exist, I came across a mob scene that would become a touchstone in my writing of this book. A crowd had gathered to attend the meeting's yearly Food and the Brain session, featuring a summary of the latest data on how what we eat influences our mental health. I angled my way through the crowd to witness Columbia University psychiatrist Drew Ramsey handing out raw oysters.

Watching a queue of shrinks dribble brine down their beards before dawn was unsettling. The rest of the presentation was anything but. For three hours, Ramsey and psychiatrist Emily Deans expounded on the ways our diet influences our brain health and mental state, and how through millions of years of evolution, what we ate, and how we sourced and prepared our food, was critical in shaping our most fantastic organ. As an MD turned health and science journalist, I knew from years of medical training and neuroscience research that certain dietary patterns were good or bad for brain health, but I'd never really considered them from an evolutionary standpoint. The talk left me curious about the influences that shaped our brain's development.

Omnivorism was crucial to our survival. A varied diet meant we could adapt to shifting food supplies brought on by climate change. If a warm spell dried up the fruits of the forest, subterranean tubers on the plains would suffice. Some believe that seafood helped save our species—when early *Homo sapiens* learned to gather and crack open shellfish along the coasts of Africa (hence Ramsey's oysters). And just about all scholars on human evolution agree that if it weren't for meat, which supported our cranial expansion, we'd all still be sitting around primitive campfires with much smaller brains than we're accustomed to. Becoming carnivorous was among the most dramatic plot twists in the human story—it helped balloon our brain size. After learning to scavenge and to hunt gazelles on the African savanna, we eventually harnessed fire to cook our spoils, further steering our mental journey with preserved, more digestible food.

I'm not advocating for eating meat, this is just what happened. What we ate was one of many influences on the human saga. Equally important was how we obtained and processed food; how we hunted, foraged, and made primitive tools for protection and butchery; how shifts in climate affected our diet, lifestyles, and physiques; how we socialized and communicated. All these factors came together through millions of years of Darwinian evolution to help shape our species and our brain.

I remember back in medical school, a few months into anatomy lab, it was time to touch a human brain for the first time. By this point, medical students are typically hardened to formaldehyde's nasal singe and the existential reckoning that comes from spending months in a room full of dead bodies. It's part of the deal. So at twenty-two years old, beholding what looked liked three pounds of a very unappetizing Jell-O variety, I was certain it would be part of my future (but as far as the University of Virginia School of Medicine knows, not something I would ever sneak visiting friends into the lab to see). I was struck by the notion that our behaviors, personalities, and conscious existence all come from a wrinkled orb of brownish mush. All of *this* comes from *that*.

The story of the human brain is meandering. It starts small, with simple, single-celled microbes and a host of weird sea creatures evolving early cell-to-cell communication, a harbinger of the neurons that would later coalesce into our nervous system. Through a branching evolutionary tree of wormy things, fish, reptiles, mammals, and monkeys, we eventually get to the apes, which branched off from other primates around twenty-five million years ago. Millions of years after that, our ape ancestors split from those of modern chimpanzees, which are, along with bonobos, our closest relatives. In the millennia that followed, many human-like species (any species belonging to the genus *Homo*) flourished. Yet *Homo sapiens* is the only one that remains. Through some combination of happenstance and ancestral adaptation, we endured when other humans did not. We weren't the strongest species on the African plains. Nor the fastest. It was our large, complex brain that kept us alive and, for better or worse, allowed us to influence the fate of the planet like no species ever before.

By analyzing genomes, crafting ancient tools, and studying ape behavior, researchers from a variety of disciplines are illuminating the story of the human psyche. It's an eonic tale of cognition that begs the question of where our brain is going. Genetic engineering technologies like CRISPR now allow scientists to literally edit our genomes in the interest of deleting detrimental genes and inserting desirable ones. For the first time in history, we will be able to artificially evolve our genetic code with precision. Some believe environmental factors like dietary patterns, chemical exposures, and the influence of technology will alter our genomes through a concept called epigenetics. It's the idea that environmental circumstances through life can change our chromosomes without actually changing the sequence of our genetic code—and that those changes are passed along to our children. Others feel that such worries are moot; that long before we raise an army of genetically engineered super babies or succumb to a cognitive breakdown from too many Doritos, we'll have run ourselves into extinction through some combination of conflict, climate change, and artificial intelligence.

As far as we know, the human brain is the most complex collection of matter in the universe. That's not to say we're more important or *better* than any other species.

As British biologists Charles Darwin and Alfred Wallace enlightened us: species arise and evolve as inherited traits influence their ability to survive and reproduce in their environment. Genes allowing an organism to last long enough to pass along their genes live on. By sheer numbers, bacteria win the evolutionary race, with a population of 5 million trillion trillion, or 5,000,000,000,000,000,000,000,000,000,000. Ants also outnumber us, as do krill. And as science through the years has shown us, many cognitive abilities long considered uniquely human exist as rudiments in other animals, especially our ape cousins. Evolution is not a progression toward complexity or intelligence with humans in the lead. It's all of us doing the best we can in our given situation.

Our brain is exceptional in so many ways both good and bad. It's benevolent. It's cruel. It's the only known entity that can think about, and operate on, itself.

In this book I'm in no way attempting a comprehensive survey of the scientific literature on human brain evolution. I'm instead trying to bring together leading theories from disparate fields, such as neurobiology, evolutionary biology, anthropology, and the budding discipline of nutritional psychiatry, in the story of how the human brain arose and evolved since the dawn of life on Earth and where it might be going as neuroscience and technology barrel ahead through the twenty-first century.

In doing so, I trace our brain back to its origins: to a prehistoric body of water, the appearance of DNA, and a coastal raw bar that may have helped save our species from extinction.

This is the story of our big, awesomely complex brain and how it got here.

PART ONE

THE ANIMAL COLLECTIVE

Homo sapiens *does its best to forget the fact,*
but it is an animal.

—YUVAL NOAH HARARI

CHAPTER ONE

VERY APE

At the San Diego Zoo in California there is a lumbering, 450-pound gorilla named Paul Donn.

According to his profile, Donn is "handsome, charismatic, and flirtatious." He's a highlight on the Monkey Trail, a winding, elevated footpath that features a variety of largely endangered primates.

In 2017, I biked to the famed zoo from my hotel to spend an afternoon observing ape behavior (and also, unwittingly, that of the entire San Diego County school system). It's well known that we share most of our DNA with other apes, and as a result, many traits and behaviors, but spend just a few minutes on the Monkey Trail, and the striking overlap in how we navigate the world takes on a more visceral meaning—fossils, behavioral studies, and DNA sequences aren't necessary to appreciate that we are all close cousins (though all three have helped confirm the fact).

When I arrived at Paul Donn's enclosure, he sat alone on his haunches, fixated on the head of iceberg lettuce a caretaker held above him. Donn nabbed the green orb with two hands as it fell and, with thick yet nimble gorilla fingers, calmly plucked off the leaves, eating them one by one. Next he walked over to the glass where I stood and picked up a small, capped cardboard container. He sat down, unscrewed the top with ease, and extracted some unidentifiable gorilla morsel. Donn wasn't just demonstrating the basic animal instinct to find food. There

Paul Donn enjoys a leafy snack.

was something more thoughtful and intentional to his behavior. Something, in a way, human.

Over at the bonobo house, a three-year-old named Belle jumped on her mother's back, picked something off mom's butt, and entered into a manic episode of double high-fives before swinging away on a series of ropes. Bonobos are close cousins to chimpanzees. They're a bit slimmer and more elegant in appearance than chimps (despite sporting the oddly parted, matted-down hair of a Nixon-era car salesman). Anthropologists describe them as *gracile*.

Like chimps, bonobos have very distinct personalities. Placards lining the Monkey Trail describe Belle as independent, her mom, Lisa, as respected and stoic, and their sixteen-year-old enclosure-mate, Vic, as friendly and mild-mannered. Despite the zoo's signage, neither Belle, nor Lisa, nor Vic, nor Paul Donn are considered monkeys. They are apes, a family of primates that branched from monkeys twenty-five million years ago into the lesser apes, which include gibbons and siamangs, and the great apes, or hominids, the big-brained ape species that include orangutans, gorillas, chimpanzees, bonobos, and humans.

Much of what scientists know about the origins of human evolution and behavior has come from observing and studying other apes and primates, whose brains are the closest evolutionary correlates to our own. We are eerily similar in our cognitive abilities, demeanors, and complicated social lives. We even share unique propensities for both altruism and coordinated brutality. The simple act of observing other apes—even through the artifice of a zoo—can put quite a dent in our self-regard as an exceptional species.

Beginning in the early twentieth century, modern scientific techniques gradually confirmed the evolutionary relationships that ape anatomy suggested. Humans split from a shared ancestor of chimpanzees and bonobos around seven million years ago, making them our closest biological relatives. Both species are genetically closer to humans than they are even to gorillas. A few weeks after my visit to California, I spoke with Janice McNernie, lead keeper of primates at San Diego Zoo Global. McNernie has over a decade of zoo-handling experience and is now working with primates exclusively. "People can recognize what they see as 'human-like' behaviors within minutes of just watching bonobos interact," she says.

More so than in most animals, spending time with bonobos reveals in them an array of behaviors and interactions they typically reserve for their own kind. Feeding, playing, and other forms of positive stimulation gradually lessen their wariness of humans—they open up, comfortable engaging in the more complex social interactions and relationships they exhibit in the wild. McNernie has witnessed their very human-like excitement over the birth of a new baby and their collective bereavement at the death of an elder. Though she's careful not to over-anthropomorphize, she feels our commonalities can't be denied. "We all descended from a common ancestor. It makes sense that we would have a genetic predisposition to share similar behaviors."

As I'll get to in more detail, bonobos and chimpanzees in particular each have distinct reputations that many primatologists feel reflect human psychological evolution. The general stereotype is that chimps are violent maniacs and bonobos their more peaceful, agreeable cousins;

thanks to our common ancestry, we wound up sharing certain traits with both species.

This idea is oversimplified, but not entirely inaccurate. Research shows that not only will bonobos share food with their friends and family, they will also do so with unfamiliar bonobo groups (Tan, 2013). This willingness to help outsiders—or xenophilia—is nearly unheard of in any species besides *Homo sapiens*. In contrast to chimps and most other primates, bonobos live in matriarchal societies, meaning females run the show. The women hold much of the social power and band together to keep rotten males in check. They help stabilize the social circle through frequent sexual activity. In turn, the typically more violent sex—men—maintain a surprising civility with each other (though they are still far more violent than human males).

"That's what most people know about bonobos: they have a lot of sex," Duke primatologist Brian Hare commented in a New York Times interview. But Hare feels that's not what makes them interesting. "The No. 1 reason they are interesting is that they don't kill each other."

Unlike bonobos, chimpanzees share our tendency to kill members of our respective species with unfortunate frequency. Aside from humans, they are the only known species that gang up to murder neighboring communities of their own kind. If some trace of matriarchal, bonobo self-control did not exist in our genomes, humans could be even more impulsive and violent than we are. Unfortunately, given our aggression and volatility, we seem to have plenty of chimp in us too.

In the 1960s, anatomist turned Berkeley anthropology professor Sherwood Washburn was a real pioneer in studying primate behavior. But it was British anthropologist Jane Goodall who popularized the field, bringing ape behavior and intelligence to the public eye. Goodall began studying the social interactions and behaviors of wild chimpanzees at Tanzania's Gombe National Park in the late 60s. It's claimed that she's the only human ever fully accepted into a chimpanzee community (though she never made it above the lowest ranking member). Through simple observation, Goodall realized that behaviors in other apes are almost certainly entangled with humanity. She noticed that, like humans, chimps each have their own developed character, whether good or bad. Some are outgoing, some shy. Some are agreeable, others intolerable monsters. They form complex emotional bonds with their families and communities; they hug; they groom; they pat each other on the back out of affection. They also make tools—chimps will strip a leaf off a branch and plunge it into an insect hole until it's crawling with a meal.

Goodall's observations were the first popular glimpse into the ape psyche. Today we can search YouTube. Apes draw. Grieve. Play games. Play keyboard with Peter Gabriel. Play bass with Flea. There's a chimp at a dubious zoo in North Korea who smokes twenty cigarettes a day (that she can light herself). As I witnessed, they can deftly disassemble a head of iceberg lettuce.

Much like Paul Donn and the bonobos in San Diego, apes are highly communicative, both vocally and gesturally, a trait mirroring not only language but also human social qualities like empathy. As I'll get to, research into how nonhuman primates perceive and interact with others has helped us understand how our brains and highly social existence came to be.

Humans aren't direct descendants of chimpanzees, or of any other extant ape species. Instead we share an ancestor. Chimps living today have also been evolving for the seven million years since our split and are no doubt different from their late Miocene predecessors—as we are from ours. So despite many similarities, our brains are still unique from those of other apes, especially when it comes to capacity for higher thought. We're better problem-solvers and have far more developed senses of emotion and self-awareness. But generally speaking, many qualities we long assumed were uniquely human appear to exist to a degree in other apes, some maintained, some muted, and some enhanced in our genomes through millennia.

When we talk about natural selection, we're primarily talking about changes to our genome that helped us adapt, survive, and reproduce. When chimpanzees join together to whack a neighboring camp, it's a safe bet this behavior is at least partially due to genetic mutations picked up through evolution that gave them some advantage. Further along, I'll dig deeper into the genetics of human brain evolution and how certain genetic variations differentiated our species.

The pop-science adage holds that we share 98.5 percent of our DNA with chimpanzees. Some estimates have our genomic overlap at closer to 95 percent. Either way, we are biologically very similar. The proteins in our bodies are nearly identical, as are our nucleotides, the code or letters that build our DNA. But considering we share 90 percent of our genes with mice (and 50 percent with bananas!) it's clear our gene sequence isn't the only factor behind our commonalities.

Equally, if not more, influential to our form and function is how our environment interacts with our genome and how our genes are regulated. Genes are the blueprint coding for the proteins that run our bodies and biological functions. But 99 percent of our DNA doesn't code for proteins at all. Most of our genetic material either just sits there inert, or, instead, controls other stretches of DNA, turning certain genes on or off.

Scientists are in the early stages of understanding how activating or suppressing particular genes at specific times may explain the variation between species with highly similar genomes. One major evolutionary driver of diversity between species is gene duplication. Genes duplicated along a string of DNA allow evolution to safely experiment with change. If a duplicate copy of a gene mutates in a way that confers a new skill or beneficial trait, then great. If a mutation renders the gene detrimental or nonfunctional, there's a backup copy to compensate.

Genetic research shows that since splitting with chimpanzees, the human lineage has acquired almost 700 duplicated genes that chimps don't have, many of them involved in brain function. We've also lost eighty-six duplicates that chimps still do have. Researchers are gradually pinpointing specific genetic influences that contributed—and still contribute—to various brain functions. They can also point to moments in our natural history when specific mutations empowered our brains and sent us down our highly cognitive course.

There is no question that ape research has clued us in to the roots of our evolution and brain function. But around the world it is slowly coming to an end, as activists convince policymakers that apes are far too intelligent and aware to be experimental subjects. In 2013, the NIH released a report calling for the end of most chimp research supported by the federal government. That same year, both the House and Senate passed a bill to expand funding for ape sanctuaries around the country. Since then, much like many humans, hundreds of chimps and orangutans have retired to places like Florida, where initiatives like the Center For Great Apes provide a home for former research apes and those rescued from sketchy carnival pasts. On a recent visit with my wife and in-laws, we witnessed a young female chimpanzee hoot at my mother-in-law while running an index finger across her lips. "At some point, someone taught her this as a sign for women," said our guide. A 2017 *New York Times* article recounted the arrival of six former research chimps to another ape sanctuary, the Project Chimps preserve in Georgia. "The group's relief and happiness was so infectious that all the humans smiled. The chimps lip-smacked and held one another's genitals," wrote reporter James Gorman.

The attempt to understand the human brain is among the most exciting scientific pursuits of our time—scientists now study brain genomics, map neurocircuitry, and literally grow brains from scratch in small plastic dishes. Observing ape behavior will no doubt continue in zoos, sanctuaries, and wild ape habitats. But the days of monkeying with captive apes in the interest of science are, thankfully, over.

Back at the San Diego Zoo's bonobo enclosure, Belle came close to the viewing window, enamored with the group of us watching her. I put my finger to the glass and she did the same—like the iconic moment from *E. T.* Her eyes were wide, full of expression. Her smile, all teeth. She traced my finger as I moved it in circles across the glass. After a minute or so, she got bored and moved on.

CHAPTER TWO

LIFE FROM NO LIFE

Sometime around 150,000 years ago, *Homo sapiens* nearly went extinct.

As our gorilla, chimpanzee, and bonobo cousins scampered around as usual, and as other human species like Neanderthals and Denisovans thrived across modern-day Europe, Asia, and the Middle East, our species dwindled to just a few thousand people, eking it out in the caves of southern Africa. These were hard times, the result of a changing climate and less reliable food. But our savvy brains saved us.

Fossil records suggest early humans adapted and found new ways to survive, like tracking tides to access nutrient-rich shellfish beds. By this point, we could start fires, cook our food, and use our smarts and creativity to get by. Before long, we pulled ourselves back from the brink of extinction and migrated around the world. Millennia later, here we are.

Evolutionary forces and adaptations allowed our genus, *Homo*, to survive for over two million years, enduring scarcity and multiple brushes with being wiped from the planet. But where did the human brain come from?

As actor and comedian Steve Carell told *Wired* in 2008, "Children are very smart, in their own stupid way. A child's brain is like a sponge, and you know how smart sponges are." The irony is that the story of the modern brain actually does begin with the sponge.

Sea sponges don't look like much. Just simple, porous collections of cells that we've used to scrub our armpits and cookware for ages. They have no organs. No nervous system. They sit inconspicuously on the ocean floor.

Yet sponges represent an evolutionary tipping point, one Aristotle recognized over 2,000 years before Darwin. Sponges were a big deal in Ancient Greece, supporting an active industry of divers. Homer mentions them in both *The Odyssey,* describing how servants "with thirsty sponge . . . rub the tables o'er" after a palace feast, and *The Iliad,* as god and blacksmith Hephaestus sponges his "brows and lusty arms." Aristotle considered these odd submarine organisms to exist at the border between plant and animal life—sessile, but with a crude ability for coordinated movement, a defining feature of life we categorize as animal. He wasn't altogether wrong. The prevailing thought among biologists is that a sponge—or something very sponge-like—was a common ancestor to all animal life on Earth today. If it's true, we owe our entire existence to these things:

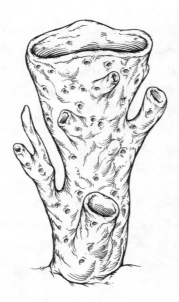 

Sponges are filter feeders—they expel water and waste and draw water in for a fresh meal (which for them consists of plankton and other microscopic morsels). Their cells use proteins to communicate with each other and keep the filtering processes going. Genetic analysis shows that these proteins are remarkably similar to those found in neurons, the cells that comprise much of our brain and nervous system and allow us to move, sense, think, and experience the strange world of consciousness (Wong, 2019).

But before we get too deep into how the brain works and how it's connected to sponges, let's explore how sponges, cells, and DNA—the helical code of all life on Earth—got here in the first place.

It all started with a chance chemical encounter in an ancient ocean.

Our planet formed from a cosmic swirl of gas and dust 4.5 billion years ago. This Hadean eon, named for Hades, Greek god of the underworld, was, as the name suggests, a hellish time. Volcanoes spewed. Asteroids pummeled. Oceans boiled away into steam.

Eventually our planet settled into a calmer scape of rock and water. And soon, through a process called abiogenesis, something approaching life arose from inorganic matter. Scientists bicker over exactly where our primordial Lincoln Logs were first assembled—some say at the surface of an ocean, others in deep-water geothermal vents, and still others argue for cold, icy pools—but most agree that four billion years ago, some recipe of earthly elements spontaneously assembled into self-replicating molecules resembling RNA and DNA, the two foundational molecules of life. DNA provides our genetic code, while RNA translates that code into the proteins that keep our cells and bodily functions running.

Scientists like University of Cambridge chemist John Sutherland have shown that Hadean Earth had the right climate and chemical cocktail to generate molecules called purines and pyrimidines, essential structural components of RNA and DNA. Over the phone on a very un-Hadean spring morning, Sutherland walked me through his theory on the molecular origins of being alive. He's confident that hydrogen cyanide—the same poisonous cyanide of Rasputin lore and Cold-War spy movies—was present at the starting gate of life. It's a simple molecule of one hydrogen atom, one carbon atom, and one nitrogen atom. When exposed to ultraviolet radiation and sulfur dioxide, cyanide has the right balance of electrons to spawn pyrimidines, which, along with purines, assemble into a sort of proto-RNA. DNA, the more complex molecule, most likely evolves later. The abiogenesis field has progressed to the point where common compounds like cyanide can also be converted to other molecules essential to life, like amino acids (the molecules that make up proteins) and certain components of lipids (fats) (Patel, 2015; Herschy, 2014).

"Once you see the chemistry, it's a beautiful link between these various classes of molecules that are essential to life," Sutherland beams through the phone. "We think they all helped each other along and arose somewhat simultaneously."

Researchers in the field agree that abiogenesis is explainable through chemical interaction—that life is possible from no life. What they don't agree on are the chemicals involved. Scores of other models have been proposed to explain the genesis of organic matter, some involving sulfur and iron, others zinc and clay. British biochemist and science writer Nick Lane believes carbon dioxide and hydrogen were life's chemical parents. Some scientists think early life arrived pre-formed on a meteorite from Mars during the Hadean bombardment, making us all, ancestrally speaking, Martians. It's possible that different stages of abiogenesis required different elements and environments and that everyone's right!

Determining where life first arose, and from what, doesn't answer the question of what life actually is. What does it mean to be alive? "It's really hard to say," admits Sutherland. Generally he sees the state of life as a continuum from the inorganic to the organic. Any definition of vitality requires replication and propagation to the next generation. So you could say that life arose when a single RNA-like molecule underwent a random mutation rendering it prone to replicating itself via the physical and chemical properties of matter. Another quality necessary to life was cordoning off our genes within a wall, at which point genetic entities started competing for survival and Darwinian evolution began.

Darwinian evolution is a beautifully simple concept: organisms and species change over generations, as inherited variations are selected to improve an individual's chances of survival and reproduction. Darwin called this "descent with modification." We now know it is a result of changes to our DNA; genetic mutations that render an organism more likely to survive in their environment and have offspring are passed along to the next generation. Mutations that weaken the chance of surviving and bearing children are gradually weeded out of the gene pool. Darwin called this survival of the fittest "natural selection" (the idea was also proposed independently by British naturalist and biologist Alfred Russel Wallace, a friend of Darwin's). Think of a chimpanzee swinging through a forest in search of dangling fruit. Mutations conferring long, powerful arms would come in handy and increase his or her chances of finding food, finding a mate, and making a few chimpanzee babies. Those same gene mutations in an early human traipsing the grassy plains on foot wouldn't have been that helpful. They weren't naturally selected for.

Evolution can also occur through a process called genetic drift. As populations become isolated, genomes change over time by chance rather than by selection acting on specific mutations.

As complex mammalian brains came along, there was cutthroat evolutionary pressure among many species to conserve genes involved in sensation and intelligence. But Darwinian evolution took hold long before that. It began the moment a replicating RNA molecule became sequestered from other RNA and could therefore outcompete or lose out to its peers—once free-floating, replicative molecules evolved into singular, membrane-bound units. These would come to look like cells, the basic units of life. "Selection really needs compartmentalization to prevent any positive innovation made by an RNA molecule from helping the other RNA molecules," says Sutherland. "The best way of doing that is to separate yourself from the other RNA."

It was survival of the fittest on an infinitesimal scale.

ANCIENT MICROBES

However life began. Wherever it began. It began.

It's safe to say that our existence took hold somewhere in the seas, in the form of single-celled organisms housing twines of genetic material much like RNA. We can't know exactly when the literal first life-form on Earth appeared. But thanks to gene sequencing, we can draw life's tree with some degree of accuracy and predict when the ur-organism showed up.

Early on, Earth's first critter gave rise to two of the three domains of life, Bacteria and Archaea. These are both single-celled life-forms with simple innards and free-floating rings of DNA. Fossils show they formed, by the billions, thick microbial mats that bobbed and bubbled in the oceans (not unlike your crunchy friend's kombucha starter). In one microbial lineage, DNA became sequestered inside a membrane, forming a nucleus and spawning life's third domain, Eukarya. *Eukarya* is Greek for "true nut," or "true kernel," the nut being the nucleus. At first, eukaryotes were single-celled, like archaea and bacteria. A million years later, they started joining together into Earth's first multicellular life.

Early microbes made quick use of ion channels to survive and help sense and react to their environment. Ions are atoms with an electric charge—for the most part, they're blocked from entering a cell's semi-permeable walls, which only let certain chemicals in and out at certain locations. This sets up an electrical gradient. But when something triggers an ion channel, or opening in the wall, the charged atoms rush into the cell, generating an electric current (Martinac, 2008; Nature Education, 2014). This flow of ions allowed single-celled life to power little tail-like projections that acted as motors. They could propel themselves toward food and away from toxins and other dangers, inching closer to life as we tend to think of it today. Naturalist John Muir went so far as to claim that these "invisibly small mischievous microbes" even had the ability to have fun (an idea actually supported by at least one modern theory on consciousness. But more on that later).

Around 3.5 billion years ago, a type of bacteria called cyanobacteria began converting water, carbon dioxide, and sunlight into sugar for energy—or photosynthesizing. A byproduct of photosynthesis is oxygen, which was toxic to much Earthly life at the time. As cyanobacteria pumped out more of the stuff, there was a mass-microbial extinction, a Great Oxidation Event, which opened up new ecological niches and shifted Earth's atmosphere to the oxygen-rich environment animals depend on today. Abundant oxygen would have led to an increase in oxygen free radicals, which damage DNA. This may have led selective pressure to drive the evolution of a nuclear membrane that could protect genetic material; it also probably later led to the evolution of sex, with selection acting on more efficient DNA repair mechanisms. Sex blends two genomes, creating genetic diversity—beneficial in adapting to a changing environment—and doing away with nasty mutations.

At some point billions of years ago, a single eukaryote ingested a single bacterium, which stuck around through subsequent generations as mitochondria, the little machines that generate energy in our cells. In the plant lineage, a gobbled bacteria evolved into a chloroplast, the structure in a plant cell where photosynthesis takes place. Both chloroplasts and mitochondria maintain their own small genomes, vestiges of their stints as independent organisms.

Archaea and bacteria remain microscopic single-celled organisms, but their ancestors also led to the multicellular eukaryotes from which all plant and animal life descend. Aside from some infamous bacteria like *E. coli* and *Salmonella*, most life we're familiar with, from yeast, to potatoes, to pine trees, to primates, are eukaryotes. And we've been locked in an evolutionary feud with our distant microbial cousins ever since we split. We pummel them daily with antibiotics to preserve our survival. In turn, they mutate and become resistant to our best medical efforts. In classic Darwinian fashion, they adapt to their environment and can readily fell even the strongest of species. Evolving a brain may have benefitted us, but it wasn't a requirement for evolutionary success.

WHEN LIFE GOT BIG

So . . . the sponge.

Once multicellular life bound itself together, it took on many forms, including various types of algae. Red algae came first, evolving at least 1.6 billion years ago. Its green cousin branched into plant life. To trace the natural history of the human brain, from here on out, I'll be focusing on the limb of life that gave rise to animals.

Scientists generally think of animals as multicellular eukaryotes that breathe oxygen, consume organic material, can reproduce sexually, grow from a hollow sphere of embryonic cells, and are able to move in some way. All animals on Earth today—from bugs to fish to humans— evolved from a single common ancestor hundreds of millions of years ago that was probably similar to the modern sea sponge. Given the lack of DNA and fossil evidence, it's difficult to know for sure when this creature arose. Early animals, unlike the bony species to come, didn't have skeletons to help etch their remains in ancient stone.

Yet by comparing the genomes of extant animal life in the context of what fossil evidence does exist, scientists believe the earliest animal life involved single cells coalescing into sessile communities and taking on specialized functions. The fossils tell us that at least as far back as the Ediacaran period (635 to 541 million years ago) the seafloor was brimming with strange and colorful stalk-like beings that flitted among the microbes. It was a peaceful time on Earth, with few if any predators; a period nicknamed the "Garden of Ediacara" after the biblical paradise. Ediacaran life is most associated with rangeomorphs, tall and beautiful organisms that looked a bit like ferns, gracefully waving with the ocean current in shades of pink, red, and green (at least according to science illustrators). Some experts consider them an alternate candidate for the earliest animal. Sponge anatomy, however, suggests otherwise (Knoll, 2006).

Sponges are in part made up of cells called choanocytes that filter nutrients from the water. Choanocytes look and function a lot like the closest living relative to all animals, choanoflagellates. These are

Above: A diorama in the Smithsonian National Museum of Natural History's Hall of Fossils imagines what the Ediacaran ocean might've looked like teeming with rangeomorphs, algae, and jellyfish

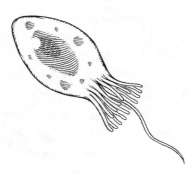

Left: Choanoflagellates are the closest living relative to all animals.

single-celled organisms with whip-like appendages that propel them through the water—picture sperm wearing beanie caps. Choanoflagellates can exist on their own or, in reaction to certain environmental conditions, aggregate into colonies in which individual cells take on specialized roles. Something similar probably took place in our shared ancestor and could explain how complex multicellular life arose in animals.

Genetic data and fossils of a sponge called *Otavia antiqua* from a 760-million-year-old rock suggest that if the sponge wasn't the first animal, it was certainly one of the first (Brain, 2012). Some researchers believe that whatever single-cellular organism first bound itself into a sponge or similar stalk must have had stem-cell like qualities, meaning it could develop into multiple different cell types.

Keep in mind that even with increasingly telling fossils and modern gene sequencing technology, tracing biology's trajectories can be an exhaustive exercise. As might be expected of a process largely based on random mutation, Darwinian evolution is a tortured mess. The tree of life is really more of shrub, all tangled branches, dead-end twigs, extinctions, new adaptations, and new species. Take the Cambrian explosion, a period beginning around 540 million years back, long considered the time when a majority of animal phyla and traits arose. Recent research shows that animal diversity actually arrived much earlier, with hard shells, movement, and hunting showing up well back into Ediacaran times. Animal evolution was more gradual than we previously thought (Fox, 2016).

However, even if life's timeline is occasionally adjusted, most biologists agree on the general progression of animal existence. After sponges, other mobile, multicellular life soon branched off our lineage, like the comb jelly, a transparent egg-shaped blob with two tentacles and tiny projections called cilia that help it swim. Combs mostly look like jellyfish, which also arose early on, as did coral, sea anemones, and flat little sea-floor creepers called placozoans. Around the same time came the first animals with a clear front, back, left, and right, earning them the name Bilateria. With bilateral beings came the body plan with which we're most familiar: mouth up front, anus 'round back, and a tubular gut in between.

Some bilaterians looked like small eels or worms. These were our ancestors—and those of all animals with a backbone—the vertebrates. Others branched into armored arthropods, which now exist as insects, arachnids, and crustaceans. Still another line split into mollusks—your snails, slugs, clams, and oysters.

Parenthetically, squid, cuttlefish, and octopus are also mollusks and have some of the most advanced brains of all invertebrates. Especially the octopus, whose brain is simply nuts. This highly intelligent animal actually has nine brains, a neuroanatomy so unique that in 2018, thirty-three scientists published a paper suggesting in all seriousness that it might be an alien—its fertilized eggs having arrived on Earth frozen in an icy meteor (Steele, 2018).

Many early animals had something approaching a nervous system. Combs and jellies move and feed using a loose network of neurons called a nerve net, so presumably their ancestors did too. Clams and other bivalve mollusks, along with their single crude foot, have a simple circuit of neurons that coordinates feeding. But a major question is how neurons themselves evolved.

Neurons, or nerve cells, look like balls of Play-Doh shaped by a caffeinated toddler. Most have one long, skinny extension called an axon on one end that sends information to neighboring neurons or muscle cells; the other end has a tree of shorter barbs known as dendrites that receive signals from axons.

When we see, taste, smell, or feel just about anything, neurons relay the information to the brain, which then processes it and decides how to respond. If the sensory input is a slice of Neapolitan pizza fresh from an 800-degree wood-burning oven, smell and taste receptors trigger sensory neurons that connect with our reward circuitry, which in turn triggers neurons in our brain's movement center to fire off a signal to our hand. We grab the slice and take another bite.

This is all made possible by the electric current generated by ion channels. Throughout our nervous system, there are supportive cells called glia that coat our axons in a fatty substance called myelin. For the most part, myelin keeps charged atoms, or ions, out of cells. Except when it doesn't. The coating has periodic breaks, or channels, that let ions in, like charged sodium, potassium, and calcium. It's a process similar to the flow of ions into other cells, only myelin isolates ionic movement to distinct locations along the axon, improving the speed and efficiency of neural signaling. This creates an electric current called an action potential that jolts down the axon, causing the release of neurotransmitters like serotonin and dopamine that communicate to adjacent cells at junctions called synapses. The collection of communicating

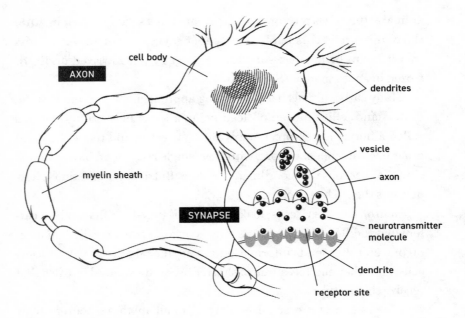

Neurons connected at a synapse

proteins in sponge cells may represent the dawn of these synapses, laying the foundation for cellular crosstalk, and for a brain to evolve.

We know neurons didn't arrive in an instant. Instead, they evolved from relatively simple elaborations on earlier cell types and traits, maybe from epithelial cells, the cells that make up our skin; or from choanocytes, the early assemblers of animal life. Sponge cells communicate using the same ion channels that bacteria have relied on for billions of years. As ions rush across their cellular membranes, chemical messengers are released to neighboring cells, including some of the same neurotransmitters the human brain and body use, like glutamate, GABA, and nitric oxide. Neuronal transmission evolved from cellular interactions that were already in place. In Bilateria, groups of axons were eventually bundled together to form well-protected nerves, the neural equivalent of those plastic bundlers that organize cords behind our televisions.

I should point out that nerves and neurons don't just communicate with each other. They also have a close evolutionary relationship with muscle, releasing neurotransmitters that cause muscles to contract and generate force and motion. The two systems evolved in parallel and enabled animals to move in new and nuanced ways—to better swim toward and wrest our mouths around food.

In a 2019 paper, German biologist Detlev Arendt wrote, "The evolution of neurons and the nervous system is one of the remaining great mysteries of animal evolution." He's open to the idea that our ancestral brain cells may have evolved independently multiple times. Comb jellies, jellyfish, and us bilateral animals all have neurons with significant differences in form and function, so it's possible there are different lineages of nerve cells, even within a single species. But there's a good chance they all evolved by co-opting the proto-synapses found in sponges. At some point between 600 and 700 million years ago, cells began communicating more effectively, giving some ancestral jellyfish-looking thing an evolutionary leg up.

CHAPTER THREE

_____

FISH, HEAD

_____

"In a sense, it takes sponges twenty minutes to sneeze."

Biologist Jordi Paps Montserrat enlightened me to this fact in an email last year. Montserrat teaches genetics and evolution at the University of Bristol in England. I'd asked him about the role of neurons in early animals. "Animals like sponges, which don't have a nervous system or muscles, can react to stimuli," he explained, "but it takes them a very long time."

Sponges use their crude cellular communication to contract their bodies, but that's the extent of their movement. Mature neurons brought more speed, more coordination, and a more exciting, if dangerous, existence to bilateral animals. In some species, selection favored those capable of quick, adept movements that helped them obtain food and avoid becoming it. Our young bilateral blueprint evolved together with an increasingly complex nervous system to better glide us through the water.

Montserrat believes the primary advantage of a bilateral body may have been ecological. Non-bilateral animals like sponges and anemones either can't move—presenting obvious disadvantages when it comes to sourcing food—or, like comb jellies and jellyfish, pulse crudely through the water hoping to bump into a meal. Jellies hunt in three dimensions and have maintained their general ball shape over millennia. Bilaterians spent more time prowling the ocean floor, a two-dimensional scape where a linear setup and more directional movement would've been advantageous. And where nutritious, easy-access carcasses settle in the sand.

If it's sponges we have to thank for our synapses, it's to bilaterians we owe our heads. Scientists imagine a mutation in an early bilaterian left a few cells at one end of its body slightly more sensitive to light or chemicals. This enhanced sensation improved its chances of survival and reproduction, thanks to more environmental awareness and better detection of risk and reward—of enemy versus energy. Light-sensitive cells begat photoreceptors, which would later gather en masse into an eye. Chemically discerning cells evolved into chemoreceptors, the cellular ancestors to our senses of taste and smell. The bilaterian mouth migrated to its consistent position up front, where food is first encountered while scuttling around. By this point, some marine worms had a precursor to a tripartite brain; meaning a brain divided into three segments—the hindbrain, midbrain, and forebrain. This is the setup for all later vertebrate brains, us included. Their hindbrain formed a rudimentary brainstem, which in more complex vertebrates helps control vital functions like breathing, heart rate, sleep, and our fight-or-flight response.

It makes evolutionary sense that over many generations natural selection would favor more and more collective sensation at one end of a long, symmetrical physique. Scientists call this cephalization; the evolution of a head with a mouth, sensory organs, and neurons clustered into a brain. Cambrian life with a clear front and back and a primitive central nervous system flourished like that space cantina in *Star Wars*, packed with bodies of every disposition.

Among the crowd were trilobites, segmented, armored animals that looked like battle-ready pill bugs, and the truly alien-looking *Anomalocaris*—or "abnormal shrimp"—which grew up to a meter long and resembled, well, huge abnormal shrimp. The branch of Bilateria we descend from were a bit more mundane—again, think something resembling a small worm—but they evolved an adaptation essential to our modern nervous system. Running down their backs was a rod of firm, cartilage-like tissue called a notochord. It gave the body structure

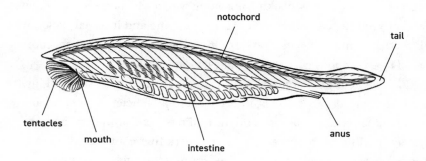

Lancelets are small fish-like filter feeders and perhaps Earth's earliest chordates.

and support, and provided protection from predators. We call any animal with a notochord at some point in their development a chordate. Lurking somewhere in the DNA of a small chordate called a lancelet is the missing link between invertebrates and vertebrates. Eventually, a notochord signaled an embryonic layer of cells to fold into a tube of neurons that would one day become the spinal cord.

In one chordate lineage, the notochord was gradually replaced during development with a spine made of cartilage, branching into cartilaginous fish, which live on today as sharks, skates, and rays. Other chordates evolved a firmer spine of bone and became the bony fish, the vast majority of fish we think of today—salmon, trout, herring, and the like. By this point, much of bilateral life had a peripheral nervous system that connected to their spinal cord and allowed for quicker, more coordinated movement.

The arrival of the skeleton was due in part to a shift in ocean chemistry. By Cambrian times, photosynthetic algae had increased oxygen levels on Earth. This gradually caused the formation of more and more limestone, a calcium source from which bone can mineralize. With an abundance of the new raw material, selection favored protecting our young sensory abilities, cooking up, along with a backbone, a protective skull around the early fish brain. Having a spine and internal skeleton allowed for a much broader range of movement among early fish, compared to arthropods with their clunky exoskeletons.

Fish had complete endoskeletons and breathed with simple gills. They had paired eyes, nostrils, internal ears—all fancier sensory organs than their lancelet forebears—and the inklings of an autonomic nervous system, part of the peripheral nerve network that controls our glands and organs. This is the maze of nerves running through our bodies that unconsciously regulates our insides, like our heart rate, breathing, and bowels. Our autonomic system is divided into two nerve networks. The sympathetic nervous system drives the fight-or-flight response to danger. It prepares us to either throw a punch or flee the scene. In both cases, our heart rate soars, our eyes open wide, and our muscles tighten to prepare for confrontation. The parasympathetic nervous system focuses on restful or enjoyable bodily functions like digestion, defecation, and sexual arousal. As the mnemonic goes, the parasympathetic system exists to rest-and-digest, or, if you prefer, feed-and-breed.

Some early vertebrates, like the hagfish, also evolved a prototype of the thalamus—a brain center that coordinates sensory information from multiple sources and generally enhances the vertebrate experience in the world—as well as a tectum, part of the brainstem involved in visual decision-making. Lamprey fish even had an early version of the basal ganglia, a central brain center that coordinates movement (its dysfunction in humans causes Parkinson's disease).

Sharks were among the first vertebrates to evolve a cerebellum, the little brain beneath the nape of our neck that coordinates balance and orientation. Early bony fish had correlates to the hippocampus, our memory center, and the amygdala, involved in fear and other emotions; both of which would form more fully when our reptilian brain came along.

Most fish were initially jawless filter feeders like lampreys—until 450 million years ago, when a pair of gills in the neck region gradually, over generations in one lineage, hardened into bone. At that point evolution went mad.

It was during the Devonian period of around 420 to 359 million years ago—the Age of the Fishes—that we vertebrates found our ruthless identity, exploiting and exterminating other species in the interest of our own survival like never before.

Previous to this, vertebrates hovered pretty low on the food chain, bested by armored arthropods. Once we had a jaw though, there was a new and brutal warfare at sea, the Edenic days of simple stalk life long gone. We evolved into awesome predators and took over the oceans using our keen senses and new bony weapon to lock in on vulnerable prey with an efficiency previously nonexistent among multicellular life-forms. As Danish paleobiologist Jakob Vinther put it to me, "We made life miserable for everyone else. We outcompeted the sea scorpions and the big-ass arthropods."

Fish don't seem that intelligent. Their brain-to-body-size ratio is on average one-fifteenth that of birds and mammals of similar size. And those blank aquarium stares don't do them any favors.

Vinther sees it differently. "Once you get a fish brain, you have the inklings of a more complex animal brain with impressive features and abilities." He notes that fish eyes work much in the same way that bird and mammal eyes do, and also that many fish are great hunters and reactors.

Once fish came to dominate the sea, there would've been huge selection pressure to be stealthy, quick, and intelligent—and for the brain and nervous system to deliver these new advantages. Fish sensory and motor systems got more complex to help outsmart and outmaneuver, and selection tinkered with new vertebrate anatomies. Visualize the progression and you'll get a picture of how aquatic life began dabbling with life on land. A few intrepid explorers made it ashore, eventually giving rise to the terrestrial vertebrate life that surrounds us today.

LAND HO

In certain parts of Pennsylvania, when the state's Department of Transportation blasts through rocky elevations to make way for highways, some very old fossil-rich rock is exposed. In the mid-2000s, at a site in eastern Pennsylvania, University of Chicago paleontologist Neil Shubin was studying an ancient, now-hardened riverbed packed with Devonian stone. As he recounts in his 2008 book *Your Inner Fish*, a shallow body of fresh water with no waves and from the right geologic age seemed the perfect place to look for creatures that may have experimented with terrestrial living.

Shubin's team unearthed a promising shoulder bone, enough to suggest his speculation was right. But with even better fossils waiting in the Canadian Arctic, he moved his dig site and made one of the biggest paleontology discoveries of the last few decades, a now-extinct fish that local Inuit elders named Tiktaalik (meaning "large freshwater fish").

It's long been accepted among scientists that in the evolutionary course of life, fish led to amphibians, and amphibians led to reptiles, the earliest fully terrestrial vertebrates. Based on genetic and general-fossil evidence, our march from sea to shore made scientific sense. Yet fossils of an actual link were elusive. Tiktaalik is that link; it's at least one of multiple similar species that made the journey to land. It lived around 375 million years ago and, like most fish, had scales and gills. Unlike most fish, it had the flat head of a reptile, the strong neck and shoulder bones of four-footed land vertebrates, and a sturdy rib cage that could support its body weight in the absence of buoyant water.

The discovery was a huge clue to how life on land began, and media attention ensued. "What the hell does that mean, my 'inner fish?'" joked Stephen Colbert. "I don't have a fish inside me."

Without the ability to retrieve oxygen from the air, migration to land would've been impossible. Like many other fish of the day, Tiktaalik had primitive lungs similar to those of its modern-day relative the lungfish. It probably relied on the awkward gulping of air that lungfish still practice when their gills aren't drawing enough oxygen from the water.

Tiktaalik is believed to be among the first vertebrate life-forms to test out life on land.

Relocating to land also required rearranging our fins. Tiktaalik had lobe-like appendages that were an anatomical cross between rayed fish fins and stumpy terrestrial legs. These transitional limbs approached our modern setup, with a bony arrangement representing the earliest known shoulder, elbow, and wrist joints. The advancement allowed Tiktaalik and other early lobe-finned fish to slosh around in shallow pools, creeks, and streams, hoisting their bodies out of the water to grab food and gulp air. Most likely, it plopped itself down in the mud and waited for an equally sluggish meal to pass by.

The most alluring aspect of spending time on land may have been safety. There would've been little competition for resources and next to no predators, save infectious microbes and disease-spreading insects (who'd already made their way to land after branching from their aquatic arthropod ancestors). Moonlighting on land would've also provided the advantage of improved reproduction rates and reliable food. Tiktaalik and the like could lay their eggs in a pool of landlocked water and expect them to sit unharmed, removed from swarms of marine predators; these pools would've also been packed with vulnerable insect prey.

Shubin thinks the migration had evolutionary implications on the animal nervous system, as the sensory environments between land and sea were so vastly different. Selective pressures on vision, detecting chemicals and sounds, and standing upright on two-dimensional ground were a literal world apart from those in the water. "These were two very different regimes," he says. Some lobe-finned fish gradually adapted to spending more time ashore. Over time, from a shared ancestor, amphibians evolved; these animals breathe air through lungs and through their skin, but they still need water to survive. The earliest amphibious creatures probably looked a lot more like Tiktaalik than the frogs and salamanders we think of today. They were hulking, a little awkward, and navigating a new ecological identity crisis.

Life on land evolved in countless directions. Over millions of years, there were changes in how we breathed, fed, got rid of our waste, and moved. Tiktaalik's club-like lobes evolved to look more like legs culminating in feet. With four limbs established, we become tetrapods, a distinction beginning with the lobed-fin fish and clearly present by the age of amphibians. At some point over 300 million years ago, a single amphibious tetrapod laid some eggs on land, and at least one hatched and survived (Pardo, 2017; Benton, 2014).

Fish and amphibians mostly lay their eggs in water. But as terrestrial life became more common, natural selection constructed a protective membrane around the vertebrate embryo. This amnion, or amniotic sac, provided a new citadel for developing young, allowing eggs to safely mature on land. Once eggs could survive out of water, amniotes—animals with an amnion encasing their embryonic and fetal youth—could roam much farther afield and populate terrestrial Earth. As vertebrate life established itself on land, one branch made the transition from amphibian to reptilian—and reptiles would flourish for millions of years as the dominant terrestrial vertebrates. From reptiles branched dinosaurs, which later gave rise to birds (which paleontologists now classify as dinosaurs). Other branches were ancestors to modern-day lizards, snakes, turtles, and crocodiles. One reptilian line evolved into synapsids and, later, therapsids, the mammal-like reptiles from which all mammalian life emerged. With the exception of insects and amphibians, most species we encounter at the zoo—including our fellow zoogoers— are either reptiles or descended from them.

The transition from amphibian to reptile to mammal brought big changes to the tetrapod nervous system. Crudely put, amphibians aren't as smart as most reptiles, which aren't as smart as most later vertebrates like birds and mammals.

Tree of Reptiles and Their Descendents

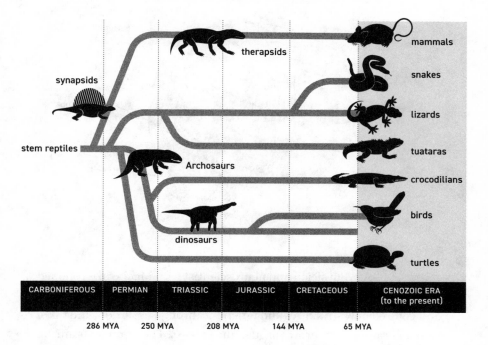

In modern humans, our central nervous system is made up of a brain and a spinal cord, with the brainstem acting as a kind of conduit between the two, contributing to vital functions like temperature regulation and the sleep-wake cycle. Behind the brainstem is the cerebellum, which oversees balance and coordination. The wrinkled, gray–Jell-O part of the brain we tend to think of is the cerebral cortex—this big, evolutionarily modern layer of our brain drives higher cognition and distinguishes great apes from most other species. The cortex is divided into four lobes. The frontal lobe controls higher thought and movement; behind it, our parietal lobe handles sensation; beneath our temples is the temporal lobe, involved in speech processing and memory; and at the back of the head is the occipital lobe, the visual center.

Major Structures of the Human Brain

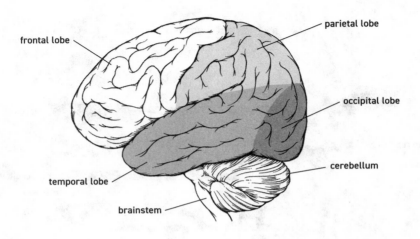

In the 1960s, physician and neuroscientist Paul MacLean developed his triune model of the brain. He proposed the vertebrate brain gradually acquired new structures and functions through evolution, and three primary brain regions arose through the process. The reptilian brain, he argued, is the most primitive among terrestrial animals, consisting of a brainstem and a cerebellum.

Next, so the theory goes, the paleomammalian (old mammal) brain came along, including our limbic system, with structures like the amygdala and the hippocampus (where long-term memories are formed). Paleo brains also have a central region called the hypothalamus that regulates metabolism, the autonomic nervous system, and components of our reward and pleasure centers. Finally, the neomammalian (new mammal) brain evolved in us higher primates. Our major upgrade was the big cerebral cortex—the reason we live our rich, hyperconscious existence. And philosophize. And drive cars. And make movies, money, and war.

As with most big ideas in science, subsequent researchers have tried to confirm, debunk, or refine the triune model, and modern science has poked some holes in it. MacLean's theory implied that each of the three major brain divisions largely controlled their own functions

Triune Brain Model

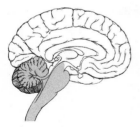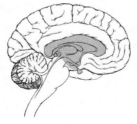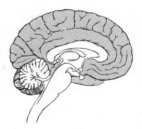

REPTILIAN BRAIN
(brainstem and cerebellum)
autopilot fight and flight

PALEOMAMMALIAN BRAIN
(limbic system)
emotions, memories,
habits, attachments

NEOMAMMALIAN BRAIN
(cortex)
language, abstract thought,
imagination, consciousness,
reasoning, rationalizing

independently: when you're feeling emotional, it's mostly your amygdala firing; when you hear a rattling sound in the woods, your brainstem alone decides whether you should fight or flee (you should flee).

Today's neuroimaging technologies, including functional MRI, can churn out brain scans that show regions from all three MacLean stages actually communicate with each other. As vertebrates evolved, visceral experiences like sex, food, and fear all took on emotion, memory, and context, as the amygdala, hippocampus, and cortex evolved to work together with primitive reptilian centers, enhancing our life experience. Reproduction was entangled with love, and feeding became a social ritual rather than a simple hunt for calories. As our cortex and memory centers enabled fear to take on more nuance, being afraid was no longer just a matter of fight or flight.

When my wife and I recently arrived home to find a towering black bear tearing through our trash can, we neither fled nor fought. Instead we stood petrified, caught up in both the primitive instinct to freeze—many predators are triggered by movement—and the more evolutionarily modern ability to reason. Our brains automatically shuffled through all the bear-related information we'd amassed through life (and had recently Googled after buying a cabin in the woods of Upstate New York). "Make your presence known, but do it calmly." "Back away slowly." "You're more likely to get hit by lightning than attacked by a bear." Our reptile brains surely wanted to run, but our neomammalian primate brain instead considered the situation more carefully. It reassured us that black bears rarely attack humans, and that they're more scared of us than we are of them. The bear walked away.

We now know that many of the brain regions included in MacLean's model had precursors in much earlier vertebrates than reptiles—like those lamprey fish with their basal ganglia. However, even if MacLean's model oversimplifies our neurobiology, it's a useful way to appreciate the course of vertebrate brain evolution. The major revision to the triune theory is that as new brain regions evolved, they did so in communication with existing brain centers and networks. In Darwinian fashion it was an ad hoc construction project that conserved older structures responsible for basic and vital functions like breathing, while allowing for additions with more neurological frills, hence all the randomly shaped regions and tangled neurocircuitry that comprise the mammalian brain.

Shubin likens brain anatomy to plumbing and wiring behind the walls of an old building. Over the years, repairs, patches, and upgrades come to resemble the architecture of a madman, a cobbling together of bygone lead pipes and corroded midcentury copper with modern PVC. It looks like a mess, but with the right plumber the water still runs. Millions of years of random genetic mutations have had a similar result in the mammal brain. Its structure isn't pretty or predictable, but it works.

THE HAIRY BEASTS

Ask five evolutionary biologists what a "mammal" is and you'll get five different answers. But only slightly different.

Everyone agrees that most mammals have fur for insulation and protection, our hearts have four chambers, and we are endotherms, meaning we generate our own heat rather than rely on the environment to regulate our body temperature like reptiles do. Perhaps the most defining mammalian characteristic is our milk-producing mammary glands ("mammal" derives from *mamma*, Latin for "breast").

After lizard babies crack through their eggshells, they're on their own in a brave new world. The therapsid lineage, on the other hand, the proto-mammals, developed a whole new means of child-rearing. As mothers began producing milk to sustain their young, their offspring could take time maturing. Instead of needing to be born ready for instant survival, mammal babies are helpless at birth, and their prolonged childhood provides more time for the environment to shape their brain. In no species is this more glaring than in humans. We're useless for years, taking in sights, sounds, and smells for nearly two decades before reaching adulthood (with many of us still psychically adrift in our thirties!). This prolonged compendium of sensory experience melds with our genetic blueprint and neural wiring to make us who we are. Eventually.

Many mammalian traits like more-vertical limbs were showing up 250 million years ago. There is a tree of increasingly mammalian branches that progresses from the reptilian synapsids to the slightly more mammalian therapsids to cynodonts, which could've passed for weasels or otters. Later, during the Jurassic period of 208 to 144 million years ago, three distinct branches of mammals split off from each other, all scurrying beneath the far more dominant dinosaurs. Some evolved into monotremes, which still laid eggs and today include only the platypus and echidna. Marsupials, which carry their young in a pouch, are another limb. Our branch is the placentals—we carry our young in a nutritive sack called a placenta.

Among the earliest evidence of a placental mammal is a fossil found in China and dating back over 160 million years. The bones suggest it was shrew-like and probably had the darting anxiety of modern chipmunks and other small animals in constant fear of being eaten. Evolutionary biologist Zhe-Xi Luo of the University of Chicago dated the remains in 2013 (Zhou).

He goes simply by "Lu," an Americanized pronunciation of his last name since, in his words, "My first name is a little hard for the Anglo-Saxon tongue to pronounce." I've called to ask him what was so special about the placental mammalian brain that it would eventually allow for the extreme intelligence seen in primates. One factor, he explains, is the sheer number of placental species compared with marsupials and monotremes. Today there are nearly 4,000 species of placentals compared with around 300 marsupial species. We simply had more rolls of the Darwinian dice. And really only two out of thousands of mammalian lineages wound up with exceptional intelligence. These are a select group of primates—monkeys, apes, humans—and the brainy sea mammals, like dolphins and orcas.

Luo explains that as certain mammal lineages evolved, the growth of their brains outpaced that of the rest of their body. Scientists measure this expansion with something called an encephalization quotient. Rather than estimating intelligence by looking at the gross size of an animal's brain relative to its body size, encephalization quotients take into account how big a species' brain is compared to how big it's expected to be based on an animal's size.

Chimpanzees have larger and more sophisticated brains than would be predicted based on the average brain volume of all mammals of similar size, like say, a large dog. If you compare skull sizes of synapsid fossils from 290 to 220 million years ago—skull size being a reasonable predictor of brain size—the brains of those that were increasingly mammalian became disproportionately larger than those of earlier non-mammalian synapsids.

Something unique about mammalian evolution and ecology drove our rapidly expanding brains.

With dinosaurs looming overhead during the day, the theory is that early mammals tended to be small and nocturnal. They would've needed heightened senses to survive, so there was selection for better night vision, smell, and hearing. Fossil findings show that, compared with early synapsids, mammals have larger olfactory bulbs, the collections of axons in our nose that transmit odors to the brain. As for hearing, in reptiles, the eardrum is connected to the inner ear by a single bone. As mammals evolved from reptiles, two jawbones were gradually reshaped and relocated to connect with that ear bone, forming a middle ear. Together the three small bones created a more sensitive conduit for sound waves, and our hearing improved. Part of the mammalian cheekbone called the petrous was also reshaped through evolution into a tube lined with little hairs that help distinguish sound frequencies. This is the cochlea, which in humans has twisted into a snail shape and in part makes up our inner ear. (Half an hour into our phone call, I am convinced Luo knows more about the petrous portion of the cheekbone than any other human.)

"We needed a larger computer to process all this information," says Luo. "The sights and sounds and smells." The new machinery came with a significant cost. Though just 2–3 percent of our body's weight, the human brain accounts for 20 percent of our energy usage—or so biology class wisdom goes. This stat dates from a single paper from the 1950s and, as far as I know, has never been reconfirmed. But the point is, the brain is one of the greediest organs in the body. It is energetically expensive.

This is where teeth became especially important. Mammals began developing different types of dentition to better process food and extract energy for their hungry brain. In the front were incisors for cutting; molars in the back evolved to mash; and sharp, pointed canines were for puncturing and biting. As Shubin wrote in *Your Inner Fish*, "Imagine eating an apple lacking your incisor teeth or, better yet, a large carrot with no molars." Our diverse diet is possible only because our distant mammalian ancestors developed a mouth with different types of teeth. We were on the road to being un-picky omnivores capable of supporting our expanding brains.

According to Luo, beyond just a Swiss Army set of chompers, the zeal with which we ate may have also helped our neuronal networks along. "Have you ever watched one of those nature shows about shrews," he asks me shortly before we hang up. "All they do is chew!"

CHAPTER FOUR

LIZARD KINGS AND LEMURS

No one is sure exactly what happened.

At the end of the Permian period, 252 million years ago, something killed off 90 percent of all species on Earth. Ninety-five percent of marine species went extinct, and over two-thirds of larger land animals perished, as did most trees. The armored Cambrian spokesperson, the trilobite, was gone for good, relegated to future tchotchke and fossil shops.

The Permian extinction was Earth's largest ever, known as the Great Dying. It marks the transition from the Permian to the Triassic geologic periods and the reboot of life on the planet. As Smithsonian Institution paleontologist Doug Erwin told *National Geographic*, "It's not easy to kill so many species. It had to be something catastrophic."

Culprits vary, but a leading theory as to what caused the Permian extinction points to volcanic eruptions. Volcanic activity at the time raised carbon dioxide emissions, which in turn raised ocean temperatures by as much as 18 degrees Fahrenheit. Some scientists believe the rise in warmth depleted the oceans of oxygen, causing a mass asphyxiation. Or it may have been the increased temperature itself that rid the world of so much life. In 2018, the *New York Times* published a piece by science writer Carl Zimmer ominously comparing our planet's current climate crisis with that of the Permian. It's title: "The Planet Has Seen Sudden Warming Before. It Wiped Out Almost Everything."

For nearly 200 million years after the Great Dying, ancestral mammals lived a humble existence. We evolved to be small and rodent-like. Many of us were arboreal, meaning we hung around in trees. A few of us burrowed in the ground like modern moles. Of main concern to us all was not getting eaten or flattened by the more cinematic dinosaurs, which dominated terrestrial animal life at the time.

The theory that most early mammals were nocturnal to help avoid the dinosaurs came about partly because many mammals today have vestigial traits for navigating darkness. Think keen senses of hearing and smell, sensitive whiskers for feeling a way through dim landscapes (whiskers are a specialized type of fur that help animals sense their physical environment), and big, often bulging eyes that can function in low light. Thanks to these adaptations, we could creep out under the cover of night to forage for bugs.

And then everything went black.

It's well accepted that sometime around sixty-six million years ago a massive asteroid slammed into Earth, causing another mass extinction that killed off nearly 75 percent of all plant and animal species, including most dinosaurs. A 100-mile-wide crater in what is now the Yucatán Peninsula helps prove it, as does a layer of sediment high in the metal iridium—which is rare on Earth but common in asteroids and meteors—found in rocks around the world and dating back sixty-six million years.

This devastating collision is called the Cretaceous–Paleogene extinction event, and is used to mark the transition from the Cretaceous to Paleogene periods. A few lineages of winged, bird-like dinosaurs— which became birds as we know them—lived on, but for the most part, dinosaurs and other large reptilian species were no more. They were unable to survive the dusty, sunless skies that would've blanketed the planet as a result of the asteroid impact, nor the tsunamis that would've careened through the Gulf of Mexico and Atlantic Ocean. It was a literal doomsday event, culminating an "impact winter," in paleontology parlance. With sunlight and radiation blocked by atmospheric debris, there was a prolonged period of cold weather. Photosynthesis in plants, algae, and plankton idled, lowering oxygen levels and limiting food for the herbivores.

Ample fossil records suggest dinosaurs and marine reptiles dominated animal life on earth for the entire 180-million-year-long Mesozoic Era. Yet not one non-avian dinosaur fossil younger than sixty-six million years has ever been found. In an evolutionary instant, they were just gone. Mammals were now free to roam.

With no dinosaurs to worry about, mammalian life could move into and exploit an expansive new ecological and evolutionary niche. All of a sudden, we could scuttle down from the trees and explore land during the day. The authors of a 2017 study compared genetic data from over 2,400 mammal species with their behavioral and sleep preferences. They found that within just 200,000 years of the dinosaurs dying out—a mere nanosecond in evolutionary terms—once-nocturnal mammals were exploring daylight en masse. This allowed for the explosion in mammal diversity that took place over the next ten million years (Maor, 2017). The first larger mammals evolved; some herbivores, others meat eaters. The first hoofed mammals, or ungulates, appeared, later branching into the ancestors of animals like tapirs, rhinoceroses, and horses. Also coming along in the early Paleogene were small placentals with more nimble hands and feet that could grasp. They had slightly more sensitive eyes that allowed for better color vision and depth perception, and eventually evolved an enlarged brain capable of a little more chicanery and cleverness. These were the primates, the order that would eventually branch off into monkeys, apes, and humans.

The dawn of primates wasn't quite as dramatic as a fish crawling to land or feathered dinosaurs branching off into birds. It was more like a small mammal that kind of looked like a rat took on some new qualities and still kind of looked like a rat.

After a few of these early transitional species, which all went extinct, the first animals with clear primate characteristics, the plesiadapiforms, arose around sixty million years ago, thriving in many parts of the world. Then came the prosimians, which today continue on as lorises, lemurs, and bushbabies, all rodentesque, but with monkey-like features.

These are strepsirhines, or "wet-nosed" primates. Around fifty million years back, we start seeing the simians, or the anthropoids, Greek for "human-like." These are generally what we call monkeys, with their expressive faces, big bulging eyes, and manic social behavior, all qualities that would serve humans quite well later on. This group are also haplorhines, or "dry-nosed" primates. We'd lost the wet noses sported by so many mammals, like dogs, cats, and deer.

Anthropoids are divided into New World monkeys, which mostly live in South and Central America—squirrel monkeys, capuchins, and howler monkeys among them—and the Old World monkeys, native to Africa and Asia. Old World monkeys are species like baboons and macaques, which would later branch into apes. Genetic data shows that Old and New World simians split well over forty million years ago, with New World monkeys soon rafting—as in literally rafting—from Africa to South America on mats of vegetation and debris. The two continents were much closer if not partially connected back then; and these rafting events are surprisingly common in evolution (Houle, 1998; Gabbatiss, 2016).

As primates evolved from shrew- and rat-like forms to be more simian, some big changes occurred. We became more omnivorous, increasingly supplementing our insect-forward diet with fruits and flowers. Flat nails replaced claws, giving primates better hand skills. And we evolved opposable thumbs and great toes that further improved our already impressive dexterity.

But perhaps most important to our success was our rapidly growing brain. Many scientists believe the monkey brain was shaped by its habitat and ecosystem, which put selective pressure on improved sensory function and intelligence to drive success and survival. Part of this was due to our return to trees.

Fossilized foot bones show that the earliest primates spent at least part of their time in the canopy. Eventually most became primarily arboreal. A friend of mine from college, John Grady, thinks about this sort of thing a lot. Grady is a biologist at Michigan State University and best known for coining the term *mesotherm* to describe animals like tuna and dinosaurs that can both regulate their own body temperature

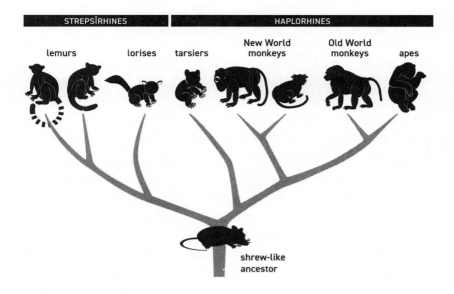

and turn to the environment for warmth. Through the years we've had countless debates about arcane bits of Darwinian minutiae, calls that typically end with one of us hanging up the phone in a huff.

Grady feels arboreal life was a coup for primates. "When you're arboreal, selection changes. Manual dexterity is beneficial—paws turn into fingers. Without the need to dig, claws can evolve into fingernails. And it's harder for predators to get you up in the trees, so coming out in the day is not so dangerous." Arboreal living also meant there was no longer a race to grow up fast before you could be eaten. You could enjoy a longer lifespan, as arboreal birds and tree squirrels do today compared to their terrestrial counterparts. "Primates could delay old age and take time maturing," says Grady. Learning became more important, as they accumulated more life experience than many other mammals.

Though we think of nocturnal mammals as having especially good eyesight, for most mammals, seeing during the day is far easier than seeing at night is even for those with the most acute nighttime vision. Primates underwent selective pressure to better navigate their day-lit surroundings and evolved to be very visual.

All animal retinas have cells called rods and cones. Rods help us see at night when light levels are low, and cones handle color vision and work best in bright light. Most mammals have just two types of cones, those sensitive to blue and green light wavelengths. This means everything they see is constructed from some combination of just two colors. After millions of years of being more active during the day—diurnal as opposed to nocturnal—some early primates evolved a third type of cone, one sensitive to red wavelengths. A more colorful three-dimensional world opened up. Selection gradually favored sight over smell or sensitivity to sound. With fewer predators to worry about, primate eyes gradually shifted to the front of their head, allowing for better depth perception.

Simians also lost a layer of eye tissue called the tapetum lucidum that reflects missed light back through the retina to improve night vision (this is what causes the glow of animal eyes we see in the forest at night). Once out during the day, monkeys and apes didn't need it; in their lineage, it was lost to evolution.

As primate eyesight improved, our senses of smell and hearing stalled compared to those of non-primate mammals. Our olfactory bulbs shrank. And though we still share plenty of genes related to a sharp sense of smell, they mostly sit dormant through the generations. "They don't necessarily cause us any harm, so they're not strongly selected against," says Grady. "We still have them, but they don't work. In primates it was all about vision." Primates wound up sensitive to the bright swirls of Van Gogh and gaudy highway billboards, whereas most other mammals only see in blurrier, impressionist pastels.

Plenty of terrestrial mammals are out throughout the day. But the ancestors of species like deer couldn't totally give up their nocturnal abilities and acute sense of smell. They're vulnerable to attack when easily visible so will forage at dawn and dusk to avoid being hunted by big cats and wolves. In the safety of the leaves, monkey adaptation could move in a new direction. As selection tinkered with our senses, our brains evolved in tow to process the deluge of new sensory information.

MACAQUES AND MOLECULAR CLOCKS

So we're living longer, and we're pretty safe in the trees, even during the day. This means there's less selection pressure to grow up and reproduce immediately. Primate evolution took advantage of the situation. We could develop for a longer period of time before becoming sexually mature, absorbing outside influences along the way. A prolonged lifespan allowed for a better memory, and an enlargement of the hippocampus. In our colorful three-dimensional world, being highly visual and smart was adaptive, so primates wound up with large visual centers, hulking frontal lobes, and a neocortex capable of higher thought and interaction with our environment. "We had the luxury of evolving intelligence," is how Grady puts it.

Another hugely important factor in primate evolution and success was being social. "Since being arboreal means you're safer from predators, it also means you're safer to socialize," he says.

When you're out during the day, sharing a tree with a bunch of monkeys, you're stuck socializing whether you want to or not. You're taking in visual information about your peers. You're interacting. Building bonds through mutual grooming. Forming enemies. Primates became very social creatures.

"Living together helped us avoid predation," says Alexandra DeCasien, a graduate student in biological anthropology at New York University who specializes in primate brain evolution. "The more bodies to look out for enemies the better." Simians could alert each other to danger by screeching their heads off or team up to mob a predator into retreat.

Communal living meant the chances of any one individual succumbing to an attack dropped with every additional member. This scenario still plays out in what English evolutionary biologist William Hamilton called the "geometry for the selfish herd," in which more dominant members of an animal population position themselves in the middle of the pack to decrease their own chances of getting picked off.

DeCasien thinks that finding food also became more effective with numbers. "Say one monkey finds a food source. The others nearby learn the news pretty quickly and benefit from it." Protecting resources would've been easier, too. Guarding an enticing branch of ripe fruit or a watering hole from interlopers is a lot easier as a group. Everyone wins.

As new social nuances formed, there was Darwinian pressure for socially intelligent brains. Primates got more politically savvy. There was runaway selection to outwit others; to rise to a higher rank. Adaptations for intelligent and social behaviors became incredibly important as one branch of Old World monkeys split off into apes, our own ancestral lineage, the lineage with the big, fancy brains.

There are a couple ways to speak about brain evolution and biology. From here on I will use brain *architecture* to refer to the different structures and areas of the brain, like our four cortical lobes. Neurocircuitry, or connectivity, describes how the brain is wired, as in, how those brain regions connect to each other through axons, dendrites, and synapses. Increased intelligence was due to both brain expansion in certain areas, like the prefrontal cortex, and changes to connectivity among brain regions. We evolved more sophisticated visual circuitry, with existing visual centers expanding. Our brains developed more and more folds, called sulci, as a result of having to pack more and more brain into a cranium. One important development was a fold called the central sulcus, which runs vertically down the side of our brains and came to separate our sensory and motor cortices. Both regions grew in size and connectivity, enhancing the sensitivity and coordination of our fingers and hands. Overall the primate brain began to take on the cortical layout and function of something approaching human-like.

As monkeys evolved, brain sizes in many simian species began to climb. This was initially confirmed in the 1960s and 70s by a number of scientists who analyzed molds, called endocasts, of primate skulls and assigned encephalization quotients to various species. In fact, it was during this time that UCLA neuroscientist Harry Jerison originally devised the encephalization formula. Rhesus macaque monkeys have an encephalization quotient of 2.1, meaning their brains are 2.1 times

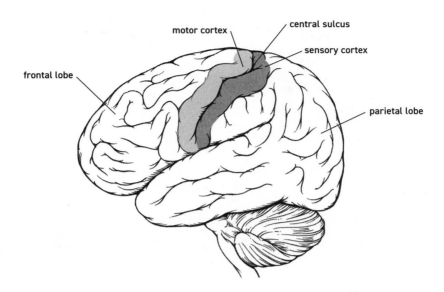

larger than would be expected based on all mammals of similar size. Chimps have a quotient of 2.5. Humans a whopping 7!

Encephalization quotients aren't an ideal measure. For one, they don't take brain complexity and wiring into consideration. Ravens and chimpanzees both have quotients of 2.5, and though pretty smart, ravens aren't chimpanzee smart. As primate brains evolved, brain structure, circuitry, and density of neurons became equally if not more important than size, as did the disproportionate growth of particular brain regions. Some scientists believe encephalization quotients overestimate values for small-bodied species and underestimate them for bigger-bodied ones. This explains why you might have read that gorillas aren't as intelligent as apes like chimpanzees.

Modern neuroimaging techniques like MRI have confirmed the exceptional nature of human brain anatomy: our cortex is far larger than expected for an animal of similar size. It has more folds and an increased proportion of white versus gray matter, suggesting more complex connections. Gray matter refers to regions of the brain composed mainly of cell bodies, whereas white matter refers to stretches of long myelin-bound axons that connect brain regions.

Encephalization Quotients

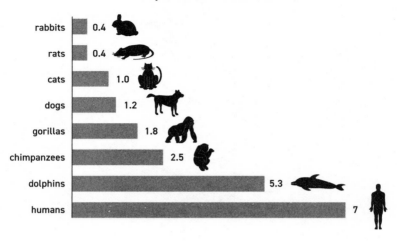

rabbits	0.4
rats	0.4
cats	1.0
dogs	1.2
gorillas	1.8
chimpanzees	2.5
dolphins	5.3
humans	7

As I mentioned earlier, the term *hominid* refers to the great apes. *Hominin* refers to modern humans and all primates that arose in the human lineage following our split with chimpanzees (all of whom, except for us, are now extinct).

In the early 1960s, scientists began piecing together how the apes relate to each other. Famed chemist Linus Pauling and biologist Émile Zuckerkandl were the first to notice that living beings have a "molecular clock." Biological molecules, like the nucleotides that comprise DNA and the amino acids that make up proteins, mutate and change at a consistent rate. By comparing differences in a particular protein or piece of DNA between species, scientists can determine when life-forms diverged through evolution, without having to rely on the often incomplete or nonexistent fossil record.

Shortly after the idea was proposed, it was bolstered by Berkeley biochemist Allan Wilson and his graduate students (Sarich, 1967). By looking at immune reactions between species, Wilson showed that humans and chimps diverged much more recently than ten to thirty million years ago, as previously thought based on available fossils. He also revealed how genetically similar our species are. Nowadays, gene-sequencing technology can sequence an entire genome and confirm how species are related to one another in about an hour. But back then, Wilson's work was revolutionary (and, expectedly, controversial).

Wilson also helped devise the equally dogma-shattering mitochondrial Eve hypothesis. Given that mitochondria were once microbes, engulfed for good as energy-providing symbiotes, they maintain a vestigial genome of their own that is passed along only through the female lineage. By comparing mitochondrial DNA in people from different populations around the world, he showed that all humans on Earth today descend from a shared maternal ancestor—our Eve—who lived in Africa 150,000 years ago. Wilson redrew the hominin family shrub.

Modern gene sequencing technologies and a growing fossil record have further refined the hominin family tree that Wilson helped sketch. We now know that among living great apes, the orangutan lineage came first, branching off around fourteen million years ago. Gorillas branched off a few million years after that, and by seven million years ago the chimp and hominin lineages had diverged. After our ancestors broke off, the chimp and bonobo lineages eventually diverged from each other somewhere between two and one million years ago.

MIOCENE MIGRATION AND CLIMATE CHANGE

Long before scientists could run DNA samples through tabletop gene sequencing machines, Darwin and others recognized the obvious kinship among primates. "If the anthropomorphous apes be admitted to form a natural sub-group . . . we may infer that some ancient member of the anthropomorphous sub-group gave birth to man," he wrote in his 1871 treatise *The Descent of Man.*

Exactly when the first ape came along is still being worked out.

"I bet if we went over there right now there'd be people debating this very question over lunch," DeCasien jokes, pointing to NYU's anthropology building. DNA and fossil evidence is, however, tightening the range of possibility.

The easiest way to tell if you're looking at a monkey or an ape isn't to greet it face to face but to scope out its rear end. Most monkeys have tails, while apes do not. Apes swing through trees with their powerful shoulders and broad chests; monkeys run from branch to branch. Apes are larger and drag the knuckles of their forelimbs. And most distinguishing of all, apes are really smart.

In 2013, anthropologists from Ohio University dug up two bones connecting a previous void in the fossil record between monkeys and apes. Digging in a Tanzanian riverbed, the team unearthed a tooth from a group of Old World monkeys and a jawbone from a new species they dubbed *Rukwapithecus fleaglei*. The findings, along with DNA sequencing data, suggest that the two groups split at least twenty-five million years ago during the Oligocene epoch (Stevens).

University of Toronto anthropology professor David Begun says that *Rukwapithecus* is the oldest known probable ape. But primatologists only have a few bones to go on. There's some degree of inference to its position on the primate tree. The earliest confirmed apes are members of the genus *Ekembo*. They lived in and around Kenya at least seventeen to twenty million years back, and more-complete fossils confirm their apehood.

Physically, ekembos mostly looked like monkeys, with limbs of equal length and a spine parallel to the ground. At this point or shortly thereafter, there were numerous ape species running around, like those in genus *Proconsul*, and later *Dryopithecus*. Though they also resembled monkeys, *Dryopithecus* species had larger brains and dangled their forelimbs like chimpanzees.

For millions of years, while the continents were still much closer together than they are today, monkeys made their way around the world using rafts and land bridges—early apes stayed mostly in Africa. That changed around seventeen million years ago, during the Miocene epoch—ancestral apes began spreading to Europe and Asia, where they thrived for another ten million years. Despite our association between apes and the forests and woodlands of Africa, for a long while there were many more apes in Eurasia.

"Some of us think that modern apes, in fact, evolved from the European apes, rather than from the small sample of African apes from the same time period," says Begun.

Most in the field, Begun included, agree that the ever-changing Miocene climate had a major influence on ape evolution and our own destiny. Throughout much of the Miocene, Earth was warm. This was probably the main reason apes could spread out into Eurasia in the first place. But thirteen million years ago, the climate started to cool. Ever-resilient monkeys ("You can drop off a monkey basically anywhere and it will probably do okay," jokes DeCasien) maintained their wide distribution, but many ape species went extinct. Others were driven back into isolated parts of the African continent, where they remain today. The climate shift may also explain why orangutans migrated south into the tropics from China at roughly the same time.

The descendants of dryopithecines were among the apes that migrated back to Africa, where they split into ancestral gorillas, chimps, bonobos, and humans. The apes that wound up surviving were the more powerful and intelligent ones, overcoming their competitive monkey cousins, climate shifts, and thousands of miles of migration.

Monkeys may have held more property than apes, but African apes of the middle Miocene evolved an intelligence that would come to change the world. It's safe to say they eventually had inklings of advanced cognition and many of the mental qualities we tend to think of as uniquely human—qualities which, in actuality, aren't unique to us at all.

CHAPTER FIVE

UPRIGHT CITIZENS

"I'm happy to talk about human brain evolution, so long as you're okay with four-letter words."

Columbia University anthropology professor Ralph Holloway swears a lot. In his eighties, he is among a handful of statesmen in the field of paleoneurobiology, the study of early hominin brain evolution using cranial casts of fossilized skulls. Last summer, Holloway invited me to visit Columbia's anthropology laboratory, where just about every afternoon—according to him—he putters.

The lab is a sea of bones: monkey skulls cram cabinets; a room-length table is invisible beneath hundreds of hominid casts and crania; two hulking gorilla skeletons stand watch from behind a glass door.

"When the hell did the primate brain become reorganized into a human-like brain?" Holloway asked, somewhat rhetorically.

"I don't know, I was hoping you could tell me," I said.

Over the next two hours Holloway enlightened me to the current thinking on how the split between our ancestors and those of modern chimpanzees came to be.

The presumed narrative holds that the divergence had a lot to do with our ancestors venturing out of the forest, if only temporarily at first. Geologists tell us that around nine million years ago much of equatorial Africa dried out. Weather patterns became more seasonal.

Holloway's Lab is a sea of bones. Endocranial casts of the fossil hominin record from 3 to 4 million years to the present and from *Australopithecus* through modern *Homo sapiens* cover the table. Gorilla skeletons procured from long-time American Museum of Natural History curator H. C. Raven's 1919 expedition to Africa stand sentry.

Lush forests shriveled into sparser woodlands and, later, grassy savannas. This put stress on apes living among trees. Many were forced to adapt to a life increasingly spent out in the open, away from protective foliage.

In the evolutionary branch that would give rise to humans, selection gradually favored traits and behaviors more suitable to terrestrial living. Erect posture and a bipedal gait would allow us to roam, while freeing our hands for other pursuits, like carrying food across long distances. Becoming increasingly social helped us forage, hunt, and defend ourselves in a dangerous new environment prowled by lions and saber-toothed cats.

Some, including biologist E. O. Wilson, have speculated that our ancestors were also in the right place at the right time. Having access to the fruits and flowers of savanna forests in addition to the tubers of the grasslands supported survival, presuming we didn't fall prey to a large feline. Selection was occurring on any number of adaptive traits, but stumbling upon a particularly giving region of reliable fruit could have helped us along.

Holloway feels we owe our trajectory and large brains to some such amalgam of factors, but admits it's impossible to say for sure which were most important. "At the moment, it's hard to know for sure what was happening as we split from chimpanzees . . . partly because the fossil evidence is almost zip," he says. "But it must've had something to do with surviving on the grasslands. That would've required intelligence."

David Begun agrees. "Why we evolved the way we did after splitting from chimps is the million-dollar question," he says. "Most likely humans became increasingly terrestrial to exploit resources on both the ground and in the trees, and to be able to move between separate patches of forest." Wilson also speculates as much. The generally accepted theory is that once we became bipedal, evolution favored certain traits adaptive for life on the plains. Those with genes conferring increased intelligence and adaptability to the local ecosystem survived.

Holloway, Begun, and others believe that for many generations after our split, the human and chimp lines would've been very similar. Nearly indistinguishable really. We would've competed for similar resources, supplementing our mostly plant-based, omnivorous diet with a little monkey meat. For a while, until our genomes diverged more significantly, we still mated with each other. But selective forces, and perhaps isolation, gradually split us apart.

Though scant, there are fossils documenting our schism with nonhuman apes. In 2001, in the arid desert of northern Chad, a team of anthropologists led by French paleontologist Michel Brunet discovered part of a skull, a few teeth, and pieces of a jaw and femur. They felt it represented a new genus of hominin and named it *Sahelanthropus tchadensis*, after Africa's Sahel region and the French spelling of Chad. Chad's president at the time nicknamed the find Toumaï, meaning "hope of life" in the local Goran language. Toumaï was originally thought to be one of the earliest known human ancestors not in the chimp lineage, but scientists now believe it may represent a common ancestor to both species. The jury is out.

Anthropologists have also amassed twenty fossils of a hominin called *Orrorin tugenensis*, which came along a million years later in what is now Kenya and, for the most part, resembled a chimpanzee. It's not clear whether *Orrorin* was a direct relative of ours, or instead a cousin whose branch fizzled out. But its skeletal anatomy suggests it could've been one of the earliest humans in our family tree, perhaps even an early bipedalist.

Once we reach five million years ago on our ancestral timeline, the fossil record blooms into a far more telling collection of bones. We start seeing species definitely considered hominins, transitional species between our early ape-like ancestors and those that increasingly resemble modern humans. The concept of a "missing link" is overly simple to describe the Darwinian tangle that is hominin evolution, but fossils reveal clear evolutionary relationships that help chart our past.

One curious genus is *Ardipithecus*, nicknamed Ardi. Fossil finds put Ardi in Ethiopia at least five million years ago, where they may have been among the first hominins to consistently explore the grasslands. Ardi's brain was about the size of a chimp's; its canine teeth intermediate between the pointy daggers of male chimps and the flattened, less-threatening canines of modern humans. The bones of its hands were those of both a climber and a sophisticate; it appears to have swung through trees yet also dabbled with bipedalism. Overall, Ardi's anatomy is bizarre. The genus may be a direct line to modern humans; or instead, like *Orrorin*, another dead-end branch on the shrub.

In 1924, anatomist and anthropologist Raymond Dart was shipped two crates of what he thought were monkey or known-ape fossils from Taung, a small town in South Africa. After weeks of carefully exposing one particular skull embedded in stone, he realized he was looking at the partial remnants of a young early human that came to be nicknamed the Taung Child. Dart believed his find was an intermediate between apes and humans, officially naming it *Australopithecus africanus*, or "southern ape from Africa," in 1925. It took years for the scientific

Tattersall's Australopith diorama at the American Museum of Natural History

establishment to buy Dart's claim, but eventually, as more fossils were uncovered, they came around. *Australopithecus*—its members colloquially known as Australopithecines or Australopiths—came to be considered the first definitively upright, bipedal hominin.

The initial proof was in the foramen magnum, a hole in the base of the skull through which the spinal cord passes and connects with the brain. In the Taung fossil, the foramen is positioned toward the front of the skull, indicating a bipedal species with a head sitting vertically atop the neck. In horizontal, knuckle-dragging apes, the foramen is closer to the back of the skull so they can lift their heads to keep their eyes facing forward.

Subsequent fossils showed that Australopiths looked exactly how you'd expect a transitional species between apes and humans to look. Picture a slightly scrawnier, less hairy chimp standing three to four feet tall on two legs. If around today, they'd be eerie reflections of ourselves on a smaller scale. Australopiths were the first hominins to use stone tools, probably for butchering, and had marginally larger brains than modern chimpanzees. Like both Ardi and other nonhuman apes, their upper body suggests they were capable climbers and still spent a good deal of time in trees. Also like apes, they had strong sexual dimorphism, meaning males were considerably larger than females. The most well-known Australopithecine is Lucy, whose skeleton (40 percent of it anyway) was discovered in 1974 in Ethiopia. She's named for The Beatles' song "Lucy in the Sky With Diamonds," which, the story goes, her excavation team blared while exhuming her fossilized remains. Lucy's discovery added to the evidence that our split with chimpanzees was more recent than previously thought.

Today, Lucy's remains float behind glass at the American Museum of Natural History in Manhattan, forty blocks south of Holloway's Columbia office. Nearby, a re-created Australopithecine couple walk side by side, the male's arm around the shoulder of his female partner. It's one of the museum's more touching dioramas.

"Why are they naked," I heard a young boy ask his father on a recent visit. "Early humans didn't always wear clothes," the dad explained, as his son laughed at prehistoric private parts.

I'd dropped by the museum to meet with anthropologist Ian Tattersall, curator emeritus of the Hall of Human Origins, and the man responsible for staging the Australopithecine couple's display. Tattersall is a hulking, understated man. On the day I visited, he sported a tie depicting what I believed to be the famed prehistoric drawings from the cave of Altamira in Spain, which seemed exactly right for the occasion. Later, in an email, he corrected me: "The tie is syncretic, with images from many caves." In any event, his presence, as we strolled labyrinthine halls, was calming. The museum's offices are connected by corridors lined with dusty lockers housing bones and bodies of every imaginable

Tree of Great Ape Evolution

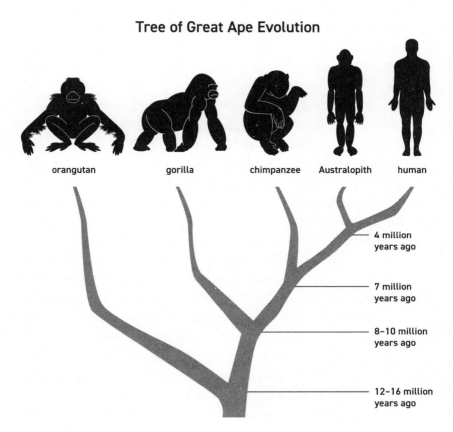

orangutan   gorilla   chimpanzee   Australopith   human

4 million years ago

7 million years ago

8–10 million years ago

12–16 million years ago

species. "We ran out of storage space, as museums do," said Tattersall, pointing to one locker out of many. "This one's full of reindeer."

Once settled in his office, Tattersall walked me through the story of *Australopithecus*.

Their genus lasted for at least two million years before giving way to early humans, and over time they branched into many species. *Australopithecus anamensis* came first and dates back four million years. The first nearly complete cranium of the species was discovered in 2016 in Ethiopia near an ancient river delta. Lucy belonged to *Australopithecus afarensis*, a slender species in our lineage that came around slightly later. The Taung Child's species, *Australopithecus africanus*, was equally slim and upright and also a close relative of *Homo*. A heartier line of early hominins were the "robust" Australopithecines, of the genus *Paranthropus*, who were probably a branch of *Australopithecus afarensis*.

In 1976, anthropologists uncovered a cluster of animal footprints fossilized in volcanic ash at the Laetoli site in Tanzania, the ancient imprints of rhinos, elephants, and antelopes. The story goes that Yale paleoanthropologist Andrew Hill, in the midst of a mock-battle with colleagues in which they hurled elephant dung at one another, dodged a clod of feces and fell. With his face planted near the ground, he noticed the prints preserved in the hardened ash. As a team of anthropologists studied the site over the next few years, a trail of footprints strikingly similar to those of modern humans were also discovered, which turned out to be those of *Australopithecus afarensis*, Lucy's species. They were made by three individuals, two of whom are the basis for Tattersall's display. The male and female were of significantly different sizes, yet their stride length was the same, implying, according to Tattersall, that they were walking in step, possibly with his arm around hers. "They somehow had to be braced against each other, and this gesture was the least loaded I could think of," he said of his curatorial decision.

The tracks were more evidence that Australopiths walked upright on two legs. Like humans, they had arches in the soles of their feet and stepped by placing one foot in front of the other. Contrast this to the ungainly stride of apes. Chimpanzees and gorillas can walk on their hind legs temporarily, but it's an awkward sight. Their pelvis is wide, forcing them to swing their legs around in a crude pivot. The motion wastes energy and isn't an ideal means of locomotion. In the event of danger, apes will typically scamper away on all fours, knuckles dragging.

It had been a long-standing question in anthropology—what came first, our big brains or our bipedalism? The findings from Laetoli and subsequent fossil and footprint discoveries prove that we were bipedal long before we were brainy.

As to why, there are plenty of theories about how being bipedal might've been advantageous. We could see farther over the savanna, and reach higher fruit in trees. Some argue that walking on two feet is more energetically efficient than lumbering around on all fours, allowing hominins to rove longer distances at less energetic cost. Also, being upright may have limited our sun exposure. Apes evolved in tropical climates and typically use the shade of trees to keep cool. Once on the plains, standing upright limited the amount of surface area getting pummeled by the African sun. The same idea may also be why we went (mostly) bald. We are by far the least hirsute primate, an adaptation resulting from our need to travel farther for food and water under hot sun. As it does for other bare mammals from hot climates—elephants, rhinoceroses, hippos—a naked physique helped us better dissipate heat.

Traipsing the plains would've led to new food sources. Eventually our free hands allowed for improved tool and weapon use, our fingers evolving to be slim and nimble and further improving on such behaviors. Some speculate that upright posture concealed the female genitalia, which in other apes are on full display during ovulation. As a result, males were befuddled as to when females were receptive to mating and more attentive to their mates. The men were more likely to stick around to maximize reproduction.

All of these ideas hold merit. But Tattersall takes pause at just how dangerous emerging from the forest would've been for early hominins. At no more than four feet tall, Australopithecines were far too shrimpy to contend with large savanna cats without confidence in their stride. "It would've been a real gamble out there," he says, claiming that the first hominins to leave the forest must have already been bipedal. "I think the only way our ancestors would have walked upright out on the plains—which would've been an extremely unusual thing to do—is if they already held their trunks upright while still in the trees and were comfortable on two legs."

Tattersall agrees with Begun that, for a long while, our migration to the plains would've been a gradual exploration: testing the waters and quickly retreating back to scattered woodland trees for cover. The Laetoli footprints go on for over eighty feet and show that the trio was crossing a wide, open plain devoid of any shelter or protection. It's thought they were heading for the Olduvai Basin a few miles away in search of water and shelter among the trees. Maybe it was worth encountering the occasional lion in the interest of a drink.

Returning to Holloway's question about the early hominin brain, Tattersall believes that, cognitively, early hominins were very similar to modern apes—complex, but nothing approaching *Homo sapiens*. A 2019 study comparing the skulls of great apes and *Australopithecus* found modern apes have a larger blood supply to the brain than the Australopiths did (Roger). Given that the brain expends 70 percent of its energy on synaptic activity, and energy delivery depends on blood flow, many in the media extrapolated the findings to mean that apes are actually more intelligent than our more closely related ancestors were. Physiologist Roger Seymour, who led the research, says the interpretation is overreaching: "Our study concerns the rate of blood flow only, and the extrapolation about intelligence is tenuous."

As *Australopithecus* evolved, there were almost certainly changes in brain structure and function. "It seems clear that stone toolmaking was invented by an Australopith," says Tattersall. "If stone toolmaking represents a cognitive leap, then later Australopiths were by definition more cognitively complex than early, non-stone-toolmaking ones." He adds that an advance like crafting tools says nothing about how Australopiths perceived the world any differently than apes. Intelligence can be defined in many ways.

It was Holloway who provided early evidence that the Australopithecine brain had evolved to be more like that of modern humans. His observations suggested that intelligence isn't just a matter of brain size, as many scientists had assumed for decades, it can also be a matter of architecture and neural reorganization.

"It must have been 1969 . . . no, that's when I got tenure . . . Jesus Christ I don't know. It must've been sometime around then." At some point in the late 1960s Holloway traveled to South Africa to visit Raymond Dart, one of his academic idols, to study Dart's famed discovery. "I saw the Taung specimen and started thinking that what lots of people were saying about brain size was all wrong—or at least not the whole story."

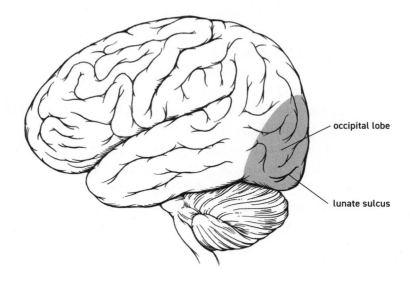

occipital lobe

lunate sulcus

There's a fold in the hominid brain called the lunate sulcus that runs through our occipital lobe, the rear part of our brain that processes vision. Dart originally proposed that in *Australopithecus*, the lunate sulcus was farther back in the brain than it is in chimpanzees, the implications being that long before our brains got really big, they'd been reorganized to resemble those of modern humans; the once hyperacute primate visual center had shrunk to make room for parts of the cortex responsible for higher thought and language. By analyzing endocasts of the Taung Child and other later fossils, Holloway confirmed that by the time *Australopithecus africanus* came along, the lunate sulcus was farther back in the brain than it is in chimpanzees.

How much we can deduce about intelligence based on endocasts is debated, as the science of analyzing sulcal patterns is a crude one. But when considering higher cognition, an obvious place to look is the prefrontal cortex, the big mass of brain sitting beneath our foreheads that allows us to plan, reason, and innovate. Holloway tells me that in early Australopithecines it's still configured like that of a chimpanzee. In later hominins it looks more like ours, proportionately much larger than in other apes. Endocast research by Florida State University anthropology professor Dean Falk, another luminary in the field, initially suggested that in Australopithecines, multiple regions of the brain, including the prefrontal cortex, had already been reorganized to look more human-like. Yet subsequent work by Falk and others using MRI scans from live chimps suggests otherwise. "Australopith sulcal patterns were completely apelike," she says.

A little over two million years ago our genus, *Homo*, branched off from *Australopithecus*, maybe from the intermediate species *Australopithecus garhi*. We went from slightly more upright chimps with a little extra going on upstairs to the genus that would come to dominate the planet. With the evolution of *Homo*, an evolutionary arms race in brain size and function was about to begin.

STUDYING ABROAD

Like so many species, we got our name from Swedish naturalist Carl Linnaeus. In 1758, he deemed us *Homo sapiens*, meaning "wise man."

Linnaeus grouped us alongside orangutans and chimpanzees in an order he called primates, as in "prime," or "first." He thought of primates as special, the highest form of animal, and among them, humans uniquely anointed. "I well know what a splendidly great difference there is [between] a man and a *bestia* when I look at them from a point of view of morality," he wrote. "Man is the animal which the Creator has seen fit to honor with such a magnificent mind and has condescended to adopt as his favorite and for which he has prepared a nobler life."

It's puzzling that Linnaeus at once believed we were special in God's creation, yet was so willing to categorize us with other animals (especially those heathen chimpanzees!). Perhaps it's because he presumed *Homo sapiens* was the only human species to have ever lived. Linnaeus believed that species were immutable, and he was simply classifying the static array of life created by God. Based on the knowledge of the time, he had no way of knowing that at various points throughout the

Pleistocene epoch of 2.5 million to 11,000 years ago, many human species roamed Africa, Asia, Europe, and Oceania. You have to wonder where other humans would have ranked on his scale of nobility had he been aware of them. The earlier humans might have really ruffled his cravat—still a bit too apish for an eighteenth-century European intellectual.

The earliest species that anthropologists classify as human-like is *Homo habilis*, which branched off over two million years back. Fossils show it as a transitional species between *Australopithecus* and later hominins. Around the same time, *Homo erectus* splintered off and may have been the first hominin to migrate out of Africa. They had a brain similar to that of modern humans and were competent travelers. Fossils found in Dmanisi, Georgia, and multiple regions of China show that *erectus* made it surprisingly far, while fossils discovered on the island of Crete attest to their crude ability to sail.

Another human species called *Homo ergaster* is debated. Some anthropologists think it was an African population of *Homo erectus*; others believe it may have been a distinct ancestral species. Around 700,000 years ago we see a big development in our evolution, the emergence of *Homo heidelbergensis*, named for Heidelberg, Germany, near which the first fossil of the species was discovered. Through the usual mess of dead-end species and evolutionary branching, it's believed that *H. heidelbergensis*, or a closely related cousin, gave rise to *Homo neanderthalensis* and *Homo denisova*—the Neanderthals and Denisovans—and, between 300,000 and 200,000 years ago, early *Homo sapiens*.

A few recent discoveries have embellished the human narrative. In 2003, on the Indonesian island Flores, researchers uncovered fossilized bones of a never-before-seen species of hominin that lived in the area until around 50,000 years ago and likely descended from *Homo erectus*. For a more modern human they were surprisingly short. Just three to four feet tall. Their discoverers named the species *Homo floresiensis*, but popular imagination quickly branded these diminutive cousins of ours "the hobbits of Flores." In 2019, another small-bodied human was found in Callao Cave in the Philippines. Archeologist Armand Mijares had uncovered evidence of early human activity there in 2003, and, inspired by the *floresiensis* discovery, he returned to see what else might lie buried in the dirt and stone. The preserved bones he and his team found are those of a previously unknown human species they called *Homo luzonensis*. Previously, it was thought the islands of Southeast Asia would've been unreachable to early humans. But it appears a few of us made it.

The other recently christened sprig on our family tree was revealed in 2013, when spelunkers discovered some bones in South Africa's Rising Star Cave. An international team of forty-seven scientists led by paleoanthropologist Lee Berger proposed that the fossils represented a previously unknown species of hominin they called *Homo naledi* ("naledi" meaning "star" in the local Sotho language). The remains date to around 250,000 years ago—making them a contemporary of early *Homo sapiens*, Neanderthals, and Denisovans.

The old idea that human evolution proceeded in some organized linear fashion is increasingly shattered with each new fossil. As humans migrated, they encountered new flora and fauna, new predators, new ecosystems, and other humans. They did the best they could to survive in new environments and, on plenty of occasions, failed. Fossil evidence shows that at least one group of Sapiens migrated out of Africa and had got as far as Greece by 210,000 years ago. But the sojourn was short-lived. They either merged with or were outcompeted by Neanderthals. Similar migrations to Israel were equally unsuccessful.

The Sapiens population that we descend from didn't leave the motherland so early. Authors of a 2019 study analyzed the mitochondrial genomes of over 1,200 indigenous southern Africans living today and determined that modern human lineage—and our mitochondrial Eve—lived around 200,000 years ago in what is now Botswana (Chan). The researchers found that a set of genes tersely named *L0* trace our lineage back to Eve and are still common in the genomes of the indigenous Khoisan people of southern Africa. All people alive on Earth are thought to descend on the maternal line back to a woman who carried the sequence, or at least to a small group of very closely related women all carrying the same genetic profile.

For 70,000 years our earliest ancestors didn't stray far from the Makgadikgadi Basin. These were lowlands circling a massive paleolake in what is now the Kalahari desert; prime territory for sourcing water and for ambushing prey doing same. At the time, the region was surrounded by arid, less hospitable terrain. Venturing too far wouldn't have been wise. But genetic evidence shows that around 130,000 our founding mothers and fathers began to roam farther afield. Around this time, according to climate models and geologic samples, increased rainfall created green corridors with more vegetation and wildlife. We could more reliably leave our lakeside home. As the climate changed, our direct ancestors spread throughout Africa, armed with a massive brain and more cognitive ability than any species ever before; some migrated out of the continent at an unprecedented pace.

SIZING UP

To measure brain size, scientists typically focus on the size of the cranium. Originally this was done by filling a skull with beads or seeds and transferring them to a measuring container. Crude for sure, but effective. Now skulls can be analyzed more precisely using CT and MRI scans. Thanks to research using a combination of old and new techniques, anthropologists have a good sense of how the hominin brain changed in size and complexity through the millennia.

Homo floresiensis had a cranial capacity of just over 400 cubic centimeters, the smallest brain of all known *Homo* species and even on the small end for Australopiths, whose brain cases ranged from 400 cc in *Australopithecus afarensis* to over 500 cc in *Australopithecus africanus*. Compare these volumes with subsequent hominin skulls and you can draw a fairly clean comparison between brain size and increasingly advanced skills and behaviors—*Homo habilis* had a brain of around 750 cc, *Homo erectus* 900, and *Homo heidelbergensis* 1,200—until you get to our hefty 1,400-cubic-centimeter melon. As for encephalization quotients—again, a measure of how big a species' brain is compared with how big it's expected to be based on the animal's size—*H. habilis* measured in at a little over 3, similar to Australopithecines and modern chimpanzees, *H. erectus* had an EQ around 4, and *H. heidelbergensis* over 5.

Through hominin evolution, the brain got proportionately larger and larger. The comparatively diminutive brain of *Homo floresiensis* is an outlier and may have been a result of isolation. A human population living a sheltered, unchanging island existence wouldn't have had the same adaptive pressures to evolve a big brain. They knew their ecosystem, predators, and food sources well. They didn't have other humans to contend with. There wasn't much environmental change for selection to act on. But the majority of humans on the mainland had to eke it out in dangerous ecosystems. A bigger, smarter brain was key to surviving and outsmarting our enemies and predators.

Neanderthals, whose ancestors split with ours over 500,000 years ago and spread across Europe and parts of Asia long before Sapiens did, are a puzzling piece of the hominin brain story. It's often said that Neanderthals had larger brains than ours, which, if we're talking about modern humans is true. Does this mean they were smarter than us? If so, why aren't Neanderthals around today, pouring seeds into our extinct skulls?

In 1856, miners came across a partial skeleton in Germany's Neander Valley. A few years later, geologist William King proposed that the bones were not human, but rather those of distinct yet related species he called *Homo neanderthalensis* (*thal* means "valley" in German). King declared that their "thoughts and desires . . . never soared beyond those of the brute," and that they would've been "incapable of moral and theositic conceptions." But it was anatomist Marcellin Boule who, over fifty years later, may have sealed our cousin's brutish reputation. A near-complete male Neanderthal skeleton had been found in La Chapelle-aux Saints in central France. Arthritic and missing most of his teeth, the poor fellow came to be called "the old man of La Chapelle." Boule reconstructed the skeleton in a slouching, ape-like pose, an image that percolated through popular culture. Science fiction writers like H. G. Wells found the concept a useful new trope; in Wells's *The Grisly Folk*, Neanderthals are strange human-like monsters to be feared.

We now know that Neanderthals were very intelligent. Like us, they'd harnessed fire to warm themselves and cook their food. They made tools from stone and bone, clothes from hides, and glue from birch tar. Their jewelry, rituals, and cave art suggest they were capable of symbolic thought, long assumed unique to modern humans. In 2018, researchers announced that cave art from three locations in Spain, previously thought to be the work of some especially creative Sapiens, had actually been painted by Neanderthals at least 65,000 years ago—twenty-thousand years before we arrived in Western Europe. Following the announcement, College of William and Mary anthropology professor Barbara J. King wrote a piece for NPR titled, "Why Won't the Old Caveman Stereotypes for Neanderthals Die?"

The Neanderthal reputation is gradually on the mend. A museum exhibit curated by Tattersall features a fully upright male carving a spear. Sitting near him, a woman scrapes an animal hide. The Neanderthal Museum in Mettmann, Germany, houses a recreation of what a Neanderthal man might look like in the modern world. He wears a blue suit and a button-up shirt; his hair is neatly combed. He looks a little apish and discreetly holds a sharp stone ax. There is something deeply arresting about seeing one of our closest human relatives, whom we almost certainly sent to extinction, dressed in modern business attire. It's both familiar and troubling in a way that a wacky chimp wearing a blazer and smoking a pipe is not.

In many ways, Sapien and Neanderthal culture was similar. But this isn't to say our brains didn't expand and evolve differently. The Neanderthal brain was elongated, like a football. Ours expanded spherically into more of a mutant volleyball. Some argue that since Neanderthals were brawnier than humans, they would've needed larger brains to control their larger bodies and muscle mass. Also, because Neanderthals had huge eyes, probably a result of living at higher latitudes with little sun and long winters, their brains would've dedicated more tissue to visual processing centers at the expense of other regions, like those responsible for socializing and cognitive skills. Even if the Neanderthal brain was enlarged, it wouldn't necessarily have been in the interest of what we traditionally think of as intellect (Pearce, 2013).

Often when comparing our species, we talk about how the modern human brain stacks up against that of ancient Neanderthals. But for 30,000 years the human brain has actually been shrinking. This could be for a variety of reasons. Holloway believes that as restructuring of brain anatomy and circuitry became increasingly important to cognition, it perhaps trumped the importance of size among more modern hominins. If we look at early Sapien human endocasts, our brains are just as big as those of Neanderthals, ranging from 1100 to 1750 cubic centimeters. At certain points in time, we both had enormous brains, far larger than humans have today. So the oft-quoted trivia about the big-brained Neanderthal comes with a caveat.

Along with changes to our brains during the Pleistocene came myriad changes to hominin physique. Our fingers got slimmer and more nimble, while our faces and jaws gradually receded. Hominin teeth changed dramatically as we became more omnivorous, our lethal canines shrinking and our molars getting flatter to grind fibrous plant foods. Fossils show a marked reduction in sexual dimorphism as males and females converged in size. In *Homo erectus*, males were 25 percent bigger than females, a larger difference than seen in humans today, but less than that of Australopiths.

A set of *erectus* footprints discovered in Kenya shows that they had heels, insteps, and toes very similar to *sapiens*, and that they walked with a gait much like ours (Bennett, 2009; Natalia, 2016). Rather than the short, more chimp-like legs of *Australopithecus*, they had long Sapien-like legs. They also had the ability to throw with accuracy.

By comparing fossils to motion captures of Harvard baseball players, the authors of a 2013 study found three essential adaptations necessary to effectively launch a projectile: a tall waist that can rotate a torso, a more rotational elbow joint, and low shoulders (Roach). Together these adaptations allowed *erectus* and later, *sapiens*, to build up elastic energy by cocking our arms and pulling off a quick release. The appearance of *erectus* two million years ago coincides with increased meat consumption and hunting. Though monkeys and chimps can crudely toss things, *erectus* and early *sapiens* could sling stones and spears with some distance and precision. Our more recent ancestors used these projectiles as long-distance weapons to help take down prey and fend off enemies.

Baseball analogies were everywhere when this study was released, as headlines declared *Homo erectus* our first starting pitcher. One may also have been our first marathon runner. In a sprint, humans were no match for many of the four-legged beasts we hunted, like antelopes and zebras. But we had stamina. We could track our prey over long distances and wait for them to tire out.

By 45,000 or so years ago, at least three big-brained hominins were roaming Africa and Eurasia. Neanderthals had reached Europe, and Denisovans Asia. Our Sapien ancestors were on both continents by now, as well as still in Africa. We lived alongside our fellow human species for thousands of years, fighting, dabbling with tools and fire, and, as the Neanderthal and Denisovan DNA stippling our genomes attests, mating with each other.

Though it's difficult to recover DNA from fossils, genetic material has been extracted from both Neanderthal and Denisovan remains. Most non-Africans have approximately 2 percent Neanderthal DNA, while the genomes of some Asian populations are up to 5 percent Denisovan, suggesting a not-insignificant degree of interspecies action.

As to why these three species arose and survived is impossible to say for sure. Unlike many scientists, Tattersall is conservative in accepting that selection acted on a few specific and essential traits that kept us around. He feels the many selective pressures and ecological influences that contributed to our successes and failures would be impossible to untangle with confidence. But he acknowledges that certain factors would've been especially influential. Being social. Symbolic communication and language. A varied diet. Adapting to a changing climate.

"All those things were necessary, but none is sufficient on its own. I think it's a very important part of our story, but I see selection as an agent to promote stability in what was a very unstable environment in Africa over the last couple million years," he says. "So adaptation to any one set of circumstances would probably have been an unwise thing."

Tattersall believes selection trimmed off the extremes, with those terribly unsuited to their environment dying out, and those more generalized and adaptable, surviving. "If there's any signal from the fossil record, it's that hominin evolution was not a matter of fine-tuning a few specific lineages over eons," he says. Instead, it was a vigorous process of evolutionary experimentation in many human species. "We've been mislead grievously by the fact that there is only one hominin species today. We tend to take it for granted and think of it as normal. It was much messier back then."

Fossils can tell us a lot about our evolutionary past. What our ancestors looked like. How they walked. What they ate. Even how our brain anatomy changed. But Holloway admits that endocasts and bits of bone only reveal so much. They show the contours of the brain's surface, yet say little about what happened in the deeper neural recesses as our brain evolved.

Though he's made his life's work studying endocasts and external brain morphology, Holloway concedes that the future of understanding our cognitive past lies in our DNA, especially given that ape research is no more. Complex qualities like cognition are likely influenced by how genes interact with our upbringing and environment. It's no easy task to determine how these influences shaped our brain, but if some combination of genes in an early hominin led to an increase in reproduction or survival, selection would have tended to conserve that genetic profile. Genomic research is already revealing how our brains may have evolved, and what neurological differences distinguish us from our ancestors, and from other apes.

There is a gene called *SRGAP2* that all mammals have. It codes for a protein that helps guide the migration of neurons during development, as well as influencing overall brain mass and the number of neurons in the cortex. Over the past three million years, the gene has duplicated three times, with its four variations sitting side by side in our chromosome 1. One of the duplicates, *SRGAP2C*, appeared 2.4 million years ago, right around the transition from *Australopithecus* to *Homo* (Dennis, 2012). The duplicated gene appears to inhibit the activity of the original *SRGAP2* gene, which kicked off an expansion of the neocortex. The new variant slowed down the brain development, allowing neurons to grow longer and build more advanced connections with one another. Over time, synaptic connections in our cerebral cortex increased. More synapses meant more neuronal communication in the brain. In turn, *Homo* became more capable, more functional, more thoughtful.

Research published in 2018 reported on another set of duplicated genes unique to humans that may also be partly responsible for our big cerebral cortex (Suzuki; Fiddes). These *NOTCH2NL* genes, as they're called, were "turned on" by a duplication around 3.5 million years ago, and subsequently three more times after that. They're expressed as our brains develop in the womb, and it's their corresponding protein that drives stem cells to develop into the neurons of our brain. More *NOTCH2NL* activity appears to slow brain development, allowing for increased cortical size. Similar genes exist in gorillas and chimpanzees, but they're nonfunctional. Among primates, only humans, Neanderthals, and Denisovans have or are known to have had the duplicated *NOTCH2NL* genes.

The body of research focused on how our brains develop, and how human-specific genes and gene-expression patterns and duplications shaped our natural history, is growing. Our cortex underwent a rapid expansion after our split with other great apes, and advancing genomic science should continue to clarify how and why our brains evolved the way they did.

"I think eventually genomics will be so advanced it will replace a lot of our older methods," says Holloway. "Genetic scientists are going to tell us that this array of genetic variants had to do with this or that trait—to the growing prefrontal cortex 800,000 years ago or to when our language centers expanded. And that's goddamn exciting."

FOREVER YOUNG'UNS

The problem with a big brain is that someone has to give birth to it.

The pelvis can only grow so much while still allowing female hominins to get around on two legs with any degree of comfort. Jogging the African plains requires an upright posture, which works a lot better with a narrower pelvic anatomy. Couple this with a bigger brain, and you have the obstetric dilemma, as scientists call it. There is no doubt that increasing brain size was important to the increasing cognitive abilities we see through mammal, primate, and hominin evolution. The primate brain is double the size expected for other mammals of similar stature. Since we split with chimpanzees seven million years ago, our brain has further tripled in size, mostly within the past two million years. This soaring growth rate led to a few evolutionary workarounds.

When we're born, our skulls have openings called fontanelles. These are those soft, impressionable areas on babies' heads. During birth they allow the bones of the cranium to overlap and slide past each other like Earth's tectonic plates; they're the reason human baby skulls are so malleable and often distorted. The fontanelle toward the front of our head stays open after we're born, allowing for the rapid brain growth we experience as infants. By around age two, it ossifies, or fills with bone; our skull plates fuse to form a fully intact, more rigid cranium. In chimpanzees and bonobos, this space is mostly closed by birth. Other great apes are incredibly smart, but they're not evolutionarily set up to grow a hulking frontal cortex in the early years of life like we are.

In 2012, Falk led a study comparing the skulls of humans, chimps, and bonobos to that of the Taung Child, hoping to determine when this delayed fontanelle closure showed up during our evolution, and by extension, when complex thought driven by an expanded frontal lobe may have evolved. Using three-dimensional computer modeling, she showed that the Taung endocast has a triangle-shaped remnant of the anterior fontanelle. This implies that the obstetrical dilemma of a big brain was solved as far back as the Australopithecines. Well before humans showed up, our ancestors had pliable, expandable skulls to make room for a big brain.

Another solution to birthing our bulky cortex was simply having children earlier in their development, when their smaller crania make for more manageable pregnancies. Most other mammals emerge from the womb ready to walk (or at least hobble) around. The fact that human babies are helpless for years is a trade-off that allows our brain to go through prolonged development outside the womb, where it's enriched by environmental influence and interactions—and free of concerns like feeding itself and avoiding predators (such worries fall to our parents). Chimpanzees and bonobos also experience a prolonged childhood, nursing for up to five years and learning to navigate their world by watching their mothers and other adults. But among the apes, it's humans who idle through an especially long period of cognitive maturation.

Even with these compensations for a hefty brain, Sapiens and Neanderthals still pushed the limits of brain expansion. Gradually, brain organization and the ways neurons connect came to be equally as important as size to evolving new cognitive skills.

There is evidence that brain architecture continued to change during the later Pleistocene. Spain's Sima de los Huesos, or "pit of bones," is one of the richest hominin fossil sites ever found. Since its discovery in the 1970s, skeletons of over thirty individuals have been exhumed from the bottom of a forty-three-foot cave shaft at the site, most likely the remains of a ritual burying. A 2014 analysis of seventeen skulls from the cave—which date to 430,000 years ago—found them to have some features associated with Neanderthals, like a prominent brow ridge, yet a more ancient cranium similar to that of *Homo erectus* (Arsuaga). They were an intermediate species, perhaps a first step toward the evolution of Neanderthals.

Endocasts show that most human species descended from a common ancestor who had cranial characteristics absent in *Australopithecus*. As Holloway and University of Wisconsin–Madison anthropologist John Hawks reported in 2018, even our small-brained cousin *Homo naledi* had undergone a remodeling of its frontal cortex resembling that of other humans. "Maybe brain size isn't all it's cracked up to be," says Hawks.

Tattersall, too, believes that changes in structure and connections became as influential as size in forming the modern human brain. "I think that's right. It really became about neurocircuitry as opposed to just size," says Tattersall. "It's a more energetically frugal way of dealing with information. It's more efficient."

One critical influence in shaping the human brain was genetic variation driving the development of "spindle neurons," or von Economo neurons—VENs for short. VENs are brain cells that allow for rapid communication in the large brains of great apes, dolphins, and elephants. They're especially prevalent in three brain regions involved in higher cognition: the anterior cingulate cortex, which drives impulse control, decision making, and morality; the insular cortex, thought to allow for consciousness and self-awareness; and the dorsolateral prefrontal cortex, involved in planning and abstract thought (Allman, 2011).

Another cell type neuroscientists love to tinker with is the pyramidal neuron, named for its triangular, pyramid-like shape. Pyramidal cells make up two-thirds of our cerebral cortex, our brain's outermost layer. In most mammals, the cortex is divided into six layers, each with distinct populations of neurons that connect with other regions of the brain; at a microscopic level, it's neuronal tiramisu. In primates, pyramidal neurons undergo extensive remodeling after birth and throughout life, shaped by our experiences. Compared with most other mammals, macaque and marmoset monkeys have a high degree of frontal pyramidal branching, chimps even more so. Humans have especially labyrinthian networks. Pyramidal neurons also comprise much of the amygdala, our brain's emotion center, in which certain populations of neurons develop at different rates. Some are already fully formed at birth. Others develop later as information pours in from circumstances outside the womb. This allows young primates to establish social and emotional connections not possible in other mammals. In a series of studies looking directly at brain tissue through a microscope, University of California, San Diego, anthropologist Katerina Semendeferi found that certain regions of the human amygdala have a disproportionately large volume and far more neurons than predicted for an ape of comparable brain size. Chimps,

Architecture of the Neocortex

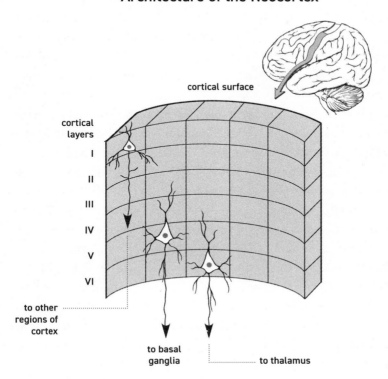

bonobos, and gorillas feel emotion, but they're not about to tear up while watching *Friends* like we do (Kaas, 2000).

By comparing the brains of great apes who have died of natural causes, Semendeferi has shown that reorganization in a number of brain regions contributed to the elevated social, cognitive, and emotional awareness that made us human. In the human cortex, there is not only an increased number of neuronal columns but also more distance between them. This increases the space available for interconnectivity between brain cells. Semendeferi also found that an area of the prefrontal cortex called Brodmann's Area 10, which is believed to be involved in memory and goal formation, grew disproportionately in humans after we split from our common ancestor with chimps.

Overall brain size was undoubtedly important to human brainpower, but Holloway, Falk, Semendeferi, and others have helped confirm that intelligence isn't that simple. The roots of wit rely on a remodeled brain with new circuitry—the enlargement of certain brain regions and the shrinking of others—all packed into what is still a proportionately enormous cranium.

Three-hundred-thousand years ago, Earth was populated by at least nine known human species. Today there is only one. The last Neanderthal walked Earth 40,000 years ago. The Denisovans lasted another 20,000 years before going extinct and ceding all of humanity to the Sapiens. Small factions of both Neanderthals and Denisovans mated and merged with our ancestors. In the case of Neanderthals, climate change may have meddled with resources and reduced fertility rates, contributing to their decline (Degioanni, 2019). But the extinction of other hominins tends to track with our arrival. The demise of Neanderthals and Denisovans came soon after they began regularly encountering *Homo sapiens*. Migrations led us to common ground, resulting, presumably, in competition, skirmish, and war. We coexisted with Neanderthals in parts of Europe for as long as 5,000 years, but eventually they were gone.

It was our large, adaptable, reorganized brain that led us to conquest. For better or worse (probably worse), we rapidly dominated Earth as no other species ever has. The other humans languished, outcompeted by our smarts, adaptability, and cooperation. We pulled this same lethal grift on many of the world's large animals too. Consistently, as *Homo sapiens* showed up, much of the megafauna went extinct. Mammoths. Giant ground sloths. A giant armadillo-like beast the size of a car called a glyptodont. Anything over ninety pounds was in our sights during the late Pleistocene. We'd become keen, intelligent hunters, unfazed at the prospect of going up against an angry six-ton mastodon. As Elizabeth Kolbert writes about in her book *The Sixth Extinction*, no species has ever led to the decimation of so much other life, and now, as the climate warms, to our planet itself.

In 2019, scientists reported on some fossilized eagle talons once strung together like beads. They were found in Spain's Foradada Cave, a known Neanderthal site dating back to 40,000 years ago. The "last necklace made by the Neanderthals," as the authors called the find, supports earlier evidence that Neanderthals had a grasp of symbolism (Rodríguez-Hidalgo). I wonder if, had their pyramidal neurons been arranged slightly differently, they'd still be around today. Had they picked up a few genetic variants expanding their prefrontal cortex or migrated to another region of the world with a different climate and more ecological options, could they have made it? Would they have pillaged Earth as we have or instead been more peaceful denizens? I'm guessing our close relatives would've had the same dueling capacities for atrocity and benevolence that we struggle with. I also wonder if Sapiens and Neanderthals could have ever lived side by side. Perhaps not. We would've competed for nearly the same stuff, with confrontation unavoidable. Assuming a slight cognitive advantage, maybe it was inevitable that one species would ultimately conquer the other.

Had things turned out differently, what would Neanderthals living in the modern world be capable of? Short of an ill-advised genetic scheme to repopulate the planet with lost human species, we'll never know.

PART TWO

OUR SOCIAL BRAIN

*Men do derive from social life much more
convenience than injury.*

—BENEDICT DE SPINOZA, *ETHICS*, 1677

CHAPTER SIX

GROOMSMEN

"The social instincts lead an animal to take pleasure in the society of its fellows," wrote Darwin in *The Descent of Man*, "to feel a certain amount of sympathy with them, and to perform various services for them."

Few would deny that humans require companions. Becoming increasingly social, and developing a richer, more efficient means of communication in symbolic language was a big part of our story. And much of our social self can be traced back to a seemingly mundane simian activity: grooming.

One primate grooms another and is, in turn, groomed. Best friends forever.

For many hours a day, monkeys and apes pluck dirt, debris, and bugs from each other's pelts. Initially the practice probably ensured that both parties stayed clean and free of disease-causing microbes and insects. It wound up a highly social exchange. Monkeys recognize when their peers are grooming one another and infer community rank based on who's cleaning whom. In macaques, females maintain lifelong relationships through grooming. The males aren't so friendly with each other, but they do groom females, especially during mating season. Grooming releases endorphins, the same euphoric opioids our bodies produce in response to pleasures like sex, drugs, and pizza. This helps calm monkey tensions after fights or conflict, stabilizing the community and reducing pain and stress. Among chimpanzees, grooming is a commodity, a bartering tool for support in a fight, or a trade for food. It's a glimpse into the ancestral mechanisms of cooperation.

Grooming relationships are flexible. Similar to how we change our behavior depending on who's around—our parents, our partner, our boss—picking fleas off each other has social nuance to it. If you're a chimp, you'll avoid grooming a friend if he or she has another friend nearby; the other friend may want in on the action and lower your chances of getting groomed. You also run the risk of your friend accepting a groom and then moving on without reciprocating. This ability to overcome a desire through what seems to be a basic form of reason and planning is rare in nonhuman animals. Like us, chimps can predict when a particular desired behavior won't go anywhere. They will also groom those of a higher social rank in the interest of future popularity and protection.

In his book *The Bonobo and the Atheist*, Emory University primatologist Frans de Waal recalls the death of a particularly beloved chimp in his research group named Amos. The ape was in rough shape, ailing from cancer and an enlarged liver but refusing to show it. He would hide in isolation for days, showing up occasionally and displaying confidence as if nothing was wrong, only to ghost again. Male chimps demonstrating any sign of vulnerability don't fair well with other males vying for alpha status. One day, Amos's health finally gave out. De Waal and his team called in a veterinarian and cracked the door to Amos's enclosure since the other chimps wanted to check in. Through the ajar door a female named Daisy groomed Amos behind his ears. She pushed wood chips through the crack, stuffing them between his back and the wall as a cushion. De Waal believes she had an awareness of how she feels better with chips between her own back and the wall and projected the desire onto her dying friend. "I am convinced that apes take the perspective of others, especially when it comes to friends in trouble," he writes. Like us, chimps sometimes appear to have a sense of compassion for each other.

Also like humans, they can be socially petty jackasses. Not unlike the paranoia of your best middle school pal switching lunch tables, or a jealous employee watching their boss play favorites, chimps are quick to thwart a budding grooming relationship if it might worsen their chances of getting groomed in the future. They keep track of threats to their social and sexual ties. It's a real communal dance that mirrors the beginnings of our own social intelligence, with its mess of emotions, attractions, and jealousies.

In the mid-1980s, Allan Wilson determined that vertebrates with a higher relative brain size had higher rates of anatomical evolution. He found evolutionary change was especially rapid in certain bird species and higher primates, and extremely quick in our genus, *Homo*; he proposed that evolution was accelerated in animals capable of "behavioral innovation and social propagation of new habits."

Having a community and being able to share information played an important role in many aspects of early humanity. A social group meant protection, sustenance, and the transfer of skills and knowledge. Accordingly, our brains grew and rewired to accommodate more collective living.

Many scholars believe becoming increasingly social played a major role in why our species won out, as it were, and came to dominate the planet. The "social brain hypothesis" holds that as we split from apes, our ancestors increasingly evolved social ways that helped us survive. Oxford University anthropologist Robin Dunbar was influential in developing the theory, originally suggesting that the more socially intelligent an animal is, the larger its brain tends to be.

The thinking has evolved to correlate brain size with the complexity of social groups and grooming networks, not necessarily absolute group size. "The issue is why do monkeys and apes have brains that are so much bigger than everyone else's, especially when we allow for body size," Dunbar asks. "The claim was made by various people that it was because they live in groups that are much more socially complex . . . I essentially provided the evidence." He adds that even among monkeys and nonhuman apes, the species with larger brains tend to live in larger groups and manage more relationships.

Humans have by far the most complex social networks of any animal, but Dunbar believes there's a cap to how much interaction we can handle. Based on the size of the human neocortex, he calculated that humans should only be able to maintain around 150 meaningful social ties—or as he puts it, the number of people you'd be comfortable grabbing a drink with at the bar. Grooming allows chimpanzees to maintain social groups of around 50 to 100 members. Dunbar believes spoken language allowed hominin social circles to grow significantly from there, but still plateau. He's shown that in contemporary Western society, people send an average of 150 Christmas cards. Other research has shown that regardless of how many Twitter followers we have, we only have consistent, meaningful interactions with 100–200 people. "Dunbar's number" appears to hold up in hunter-gatherers, industrialized societies, and even virtual social networks.

Harvard anthropologist Richard Wrangham isn't completely sold. He feels Dunbar's data is pulled from such a broad array of primate samples that assigning a specific number to our social abilities is a useless exercise. "I'm mostly agnostic here based on the evidence," he says, "but I definitely appreciate the principle that competition and cooperation within a group is cognitively taxing and may have been acted on by selection."

Catherine Hobaiter, a primatologist at the University of St. Andrews in Scotland, points out that some monkey species live in very large groups, yet tend to be less social. "The social intelligence hypothesis is a very powerful one," she says, "but there's likely to be more to what makes us human than just our ability to navigate social relationships."

Hobaiter notes that some research supports selection for spatial and technical abilities in shaping ape and hominin intelligence. It's called the "ecological hypothesis" of intelligence—the idea that an individual's ability to forage and extract nutrients from their environment is key to success and survival, for the most part, independent of social behavior.

This idea had fallen out of favor as the social brain hypothesis caught on. But a 2017 study coauthored by Alex DeCasien revived debate over whether capably exploiting our environment drove the evolution of the large primate brain. DeCasien compared brain size in over 140 primate species, including monkeys, apes, lorises, and lemurs, with their consumption of fruit, leaves, and meat. She also compared brain size with group size, social organization, and mating systems. By looking at factors such as whether or not a particular primate group prefers solitary to pair living or whether they're monogamous, she and her research colleagues figured they should theoretically be able to determine if social factors contributed to the evolution of larger brains. They found dietary preferences—especially fruit consumption—were much more influential in shaping the primate brain than social life. Fruit-eating species, or frugivores, have significantly larger brains than both omnivores and foliovores (those that prefer eating leaves).

Dunbar thinks it's impossible to untangle the two theories, as selection pressures on social behavior are inevitably related to ecology. In both monkeys and apes, pursuits like finding food are often more successful in a group. Adapting to a changing environment often plays out through a social lens. Say climate change kills off a once-thriving woodland packed with caloric fruit; you're better off finding a new food source as a group, with someone pulling up an edible tuber and alerting the others. "If food-finding is what limits their ability to survive, then sharing information is what is important," says Dunbar. "If predation is the big problem, then staying together as a group is [beneficial]. The social brain hypothesis *is* an ecological hypothesis."

CAMPING GROUNDS

E. O. Wilson argues early humans had a leg up by being eusocial, meaning their groups had multiple generations living together, abided by a division of labor, and were capable of altruistic behavior. There aren't many of us eusocial beings. Aside from humans, eusociality only exists in termites, a few species of bees, and ants, on which Wilson has been called the world's leading expert (in 1990 he wrote a 732-page book about them).

Wilson observed that eusocial animals build nests, which they protect from others. In our ancestors you might call them camps, settlements that, after a day of foraging or hunting, they'd return to, gathering around a fire for warmth and cooked food. The first human camps weren't permanent. Hunter-gatherer humans would've moved on once local resources were exhausted. But this was a new hominin lifestyle. Instead of roaming all day and settling at night where they may, hominins increasingly adopted camping culture, spending evenings together and securing their nest, bringing the spoils of a forage or hunt back home to share with others. More and more, subsequent generations would stick around and help care for community elders. Fossils and archeological evidence show *Homo erectus* was building these campsites as early as one million years ago.

If you're ever in a crowd of biologists and really want to stir things up, bring up the concept of group selection. Many modern biologists believe natural selection acts on the individual—that we're each out to propagate our own DNA. Group selection, rather, allows room for selection to occur at the community level, with some individuals sacrificing their own reproductive success for the good of the collective. If seemingly altruistic behaviors like sharing food or throwing yourself into battle benefits the group, then the collective genome of that group will live on even if those of a few individuals do not. Wilson believes we evolved through both individual and group selection, and our genomes and traits are in constant conflict between cooperation and selfishness. He writes, "The human condition is an endemic turmoil rooted in the evolution processes that created us. The worst in our nature coexists with the best, and so it will ever be."

Other researchers, including evolutionary biologist Richard Dawkins, have slammed any acceptance of group selection, arguing that while organisms may occasionally exhibit altruistic acts, it's because doing so helps propagate their genetic code. In his 1976 book, *The Selfish Gene*, Dawkins argues our bodies are mere "survival machines" for our genes. Our time on Earth is fleeting, but, assuming we have children, our DNA lives on. At its extreme, the theory holds that animal behavior exists almost entirely to propagate DNA. There is always a selfish motivation lurking in our genomic wings.

It's rational to assume that by joining a group and engaging in social behaviors that result in your protection, you're acting in the interest of preserving your own genes. Philosophers have been going at it for centuries. Can humans genuinely be good people? Are virtue and altruism innate? Dawkins argues selfish genes can sometimes result in altruistic acts in organisms if they increase the likelihood of propagating those genes. There is also the concept of reciprocal altruism, in which one organism helps another in the expectation of something in return: "Sure, have some of these figs I spent all day foraging. But you'll get me back when I'm running low. I'll scratch your back . . ."

A 2017 study by Michael Tomasello, long-time director of the Max Planck Institute for Evolutionary Anthropology in Leipzig, Germany, found chimps are willing to help a peer obtain food at their own personal cost if that peer had previously helped them access food (Schmelz). It's possible that among humans and other apes, helping strangers minimizes conflict among the group, and improves each individual's chances of safety and survival. Still, humans, bonobos, and chimps all demonstrate what appears to be some degree of genuine compassion, even if we're all being foiled by our genomes.

I'm staying well clear of the evolutionary maelstrom around individual versus group selection, but Wilson's argument that a more communal life centered around campsites makes sense. When living together in more permanent settlements, humans were forced into a new model of social interaction. They had to cooperate and divide responsibility to keep the camp running smoothly. Some hunted and foraged, others stayed back to tend the fire and protect the young.

With nursing mothers increasingly bound to the camp to care for helpless, big-brained children, a new biology of parenting evolved. New moms had to rely on their mates, family members, and friends for survival. Selection favored social intelligence. One theory even holds that simply living long enough to have grandparents who could help with childcare would've been hugely advantageous to early humans. Community played an important role in so many aspects of early humanity, like acquiring food, protection, and sharing knowledge and information.

Humans are a tribal, fraternal lot. Groups gave us protection, watchmen, collaborators. Our modest campsites eventually grew into societies, as people with common languages, cultures, and religions joined into larger and larger assemblies. Wilson believes humans were genetically and cognitively set up for one day taking over the planet. We had the right preadaptations, qualities that, farther down the road, enable new skills and behaviors. Thanks to our simian past, we had opposable thumbs and our brains were primed for the development of increased social behavior. For cooperating. Creating. And eventually talking to each other.

Early human existence wasn't all fireside lounging. Our burgeoning new lifestyle would've come with the usual evolutionary competition manifested through shifting social dynamics. Determining who enjoyed a larger portion of the food and the most desirable mate would certainly have boiled over into not just the expected alpha posturing but also more nuanced political and psychological angling than that seen in other primates. Those better at reading the intentions of others would've had the advantage. They could build stronger alliances, gain trusts, and detect potential rivals. "Social intelligence was therefore always at a high premium," according to Wilson.

A darker side of huddling together in the evenings would've been the xenophobia it encouraged. As we guarded our camps, there may have been a value to insular thinking, as we protected our resources. Sadly, being suspicious of those less familiar to us is hardwired in our DNA. We're wary of Others. Outsiders. Those not part of our *in-group*, as psychologists put it.

Psychology studies show humans are more at ease with those more like themselves. We prefer familiar social and cultural circles. Whether based on race, religion, nationality, or social class, a sense of favoritism in one form or another can develop in children even before they learn to speak. It was assumed these biases primarily arise culturally via parental and community attitudes. And in some cases they no doubt do. But a 2019 study published in the *Journal of Experimental Child Psychology* found the same xenophobia in randomly assigned groups. Children between five and eight years old were more likely to play and share with children in their own group. This held true even when a child was abruptly switched to another group. In the absence of any physical or cultural distinctions, "membership" of any group at all, however arbitrary, instills allegiance and social preference. The researchers believe this was an adaptive strategy to encourage cooperation and improve chances of survival. They see it as a basis for future research into how to quell our innate tendency toward prejudice, racism, and discrimination.

Human tribalism bore out in ways brutal and benevolent. We could march together in battle or share our fire-starting skills back at camp, both potentially improving our chances of outcompeting the Other and protecting our community. In either case, our genes had a chance of living on.

THE MONKEYS ARE WATCHING TELEVISION

At some point early on in childhood, usually between three and five years of age, human children become aware that other people have thoughts, opinions, and desires unique from their own. We emerge from solipsistic youth to appreciate the intentions and perspectives of others. The concept is called theory of mind, or TOM. University of California, Santa Barbara, psychologist Michael Gazzaniga describes it as "the ability to observe behavior and then infer the unobservable mental state that is causing it."

For years, psychologists have wondered whether or not any species other than humans can understand the psychology of others, with some concluding either they don't, or no evidence exists to prove they do. Recent research suggests otherwise. An experiment by Michael Tomasello and Brian Hare found that when chimpanzees and humans compete for food, the chimps consistently approach the food via a path hidden from the human. They use what others can and cannot see to their advantage. Tomasello's work also shows chimps have at least a basic understanding of human intention. Less clear is whether they appreciate when someone else holds a false belief, or believes something that isn't true.

In 2016, anthropologists Christopher Krupenye and Fumihiro Kano made a movie and screened it for a group of chimps, bonobos, and orangutans. In the opening scene, Man One, who is dressed to appear ape-like, steals a rock from Man Two. Man One then places the rock in one of two boxes and scares Man Two away. The thief then transfers the rock to the second box. Most of us would assume that when Man Two returns he will look for the rock in the first box; we have an awareness that his knowledge of the situation is not as complete as our own. But do other apes recognize this? Using eye-tracking analysis, Krupenye and Kano found that among those apes interested in the movie in the first place, 77 percent stared at the first box. They had some awareness Man Two held a false belief and were able to predict where it would lead him.

Many psychologists and neuroscientists believe the ability to understand the perspective of others—and to mimic them—was a seed for developing more complex social qualities like sympathy and empathy. It let us appreciate what our friends and families were going through, and what schemes our enemies were hatching up. We could glimpse the motivations of others, and, when it benefitted us, copy them. Literally "aping" another's behaviors allowed for skill transfer; it also allowed parents to pass on information to younger generations. We could determine who was trustworthy and who wasn't—whom we should love, befriend, envy, and despise. All the psychic emotions and traits that make us human are owed in part to our reading the minds of others and evolving increasingly sophisticated social intelligence. By the time *Homo erectus* was sitting around the fire in a new social scene, each individual would've had a keen awareness of their campmates' perspectives. "A sharp sense of empathy can make a huge difference, and with it an ability to manipulate, to gain cooperation, and to deceive," writes Wilson.

In early hominins, communal and cultural behaviors probably accelerated selection pressure, as certain genes and genetic profiles proved advantageous in our increasingly social existence. This hasty, community-based evolution would become an especially important meter of success throughout the last few million years, during which the cranial capacity of our lineage ballooned. In 2007, Tomasello and his Max Planck colleague Esther Herrmann published a study in which chimps, orangutans, and human toddlers were administered a battery of cognitive tests assessing how they navigate their environment, both physically and socially. The physical tests were designed to look at behaviors like spatial understanding and tool use; the social component tested study subjects on how well they imitated another's solution to a problem, communicated nonverbally, and demonstrated theory-of-mind. The apes and children performed similarly on the physical challenges. But the toddlers out-performed both ape species in tasks involving social intelligence, suggesting our more attuned sense of social awareness was important to the evolving hominin brain.

This is not to say other primates aren't socially intelligent. They are.

It's more that there's a continuum in social intelligence as you move from monkey to ape to hominin. Chimps may have an awareness of what their competitors can and cannot see. But through our evolution, social awareness took off. Gazzaniga gets at this nicely with a TOM brain teaser that, at a certain point, humans had the smarts to make sense of.

Imagine a dialogue between two people that goes like this: "I know that you know that I know that you want me to go to Paris."

Easy enough.

But taking the example a few steps farther down the TOM rabbit hole becomes more difficult: "And you know I can't, and I know that you know I can't."

Regardless of when outright theory of mind came along, the neurocircuitry of social primates was in the works well before apes split off from Old World monkeys. Research by Winrich Freiwald, a neuroscience professor at The Rockefeller University in New York City, found macaques have a neural network in their brain akin to our own brain's social wiring.

In a 2017 study led by Freiwald, macaques were shown videos of various social and physical interactions while undergoing functional MRI, a neuroimaging technology that measures brain activity by detecting changes in blood flow. They watched clips of other monkeys either interacting or performing tasks on their own. The macaques also watched videos depicting physical interactions between inanimate objects.

As expected, viewing videos of monkeys doing anything at all activated brain regions involved in face and body recognition, whereas scenes of objects stimulated regions that help with object identification (Sliwa). More intriguing was the discovery of a large macaque brain network that was activated when the monkeys viewed social interactions. Watching videos of other monkeys engaging socially lit up what Freiwald calls the "social-interaction network," which includes the prefrontal and anterior cingulate cortices, both packed with spindle neurons and responsible for higher thought and feeling. The results show that the social network in the macaque brain looks a lot like the human brain's TOM circuitry.

The brain is full of redundancy, with different regions often engaged in multiple mental and physical processes. This is why it's so difficult to understand exactly how the organ works! But Freiwald's findings show the macaque social brain network has a clear function in analyzing social situations. Understanding the interactions of others is an important part of primate life, and natural selection ensured this ability would have some dedicated machinery.

Some scientists speculate our ability to understand social situations, express empathy, and communicate is in part due to our mirror neuron system. Mirror neurons are brain cells that fire not only when we carry out a certain activity or motion but also when we watch someone else do it. The theory holds that this system is essential to understanding the intent of others—and to imitating them in order to learn new skills. Freiwald and his colleagues detected mirror neuron activity when macaques viewed monkeys socializing. Yet their brains responded similarly to videos of two toys interacting. The observed brain activity might not be exclusively a response to social engagement, but instead a product of processing physical interactions of any kind.

Italian neuroscientist Giacomo Rizzolatti was the first to describe mirror neurons. Using neuroimaging, he observed neurons in the macaque brain that fire when the monkeys either grasp an object or witness a researcher do the same. Functional MRI studies show humans demonstrate similar brain activity when executing, observing, or even just imaging a particular movement or action. Since then, some scientists have argued the mirror neuron network forms the basis of empathy and appreciating someone else's perspective. However, research shows it can't be this simple.

Most mirror neuron activity occurs in brain regions not associated with empathetic responses. Empathy is a complicated state, coordinated by a wide network of neuronal centers, some of which overlap with Freiwald's social network. Also in play when we empathize or sympathize is oxytocin, the "cuddle hormone." This tiny molecule released by the pituitary gland originally evolved to support maternal-child bonding and nursing, but through generations has come to drive sexual attraction, love, and social relationships.

Neuroscientist Vilayanur Ramachandran once claimed mirror neurons shaped our civilization. Now, scientists are a little hazy on their relevance to our evolution and social cognition. Many feel they have far more to do with recognizing and mimicking actions than they do with appreciating the perspective of others. Psychologist Christian Jarrett has said mirror neurons are the most hyped concept in neuroscience, pointing out that learning can alter their activity. "It can't reasonably be claimed that mirror neurons *made us* imitate and empathize with each other, if the way we choose to behave instead dictates the way our mirror neurons work," he wrote. "[They] are fascinating but they aren't the answer to what makes us human."

Mirror neurons may very well have a role in our social equation, but in later hominins an awareness of what others are thinking—understanding their motivations, empathizing with them, sometimes not empathizing with them—evolved into a neurological labyrinth involving many interconnected reaches of the brain. It's difficult to prove one factor or another contributed to human brain evolution, but the evidence is strong that adapting to group living and social pressures was a critical step in the evolutionary march from monkey to mankind.

CHAPTER SEVEN

A HISTORY OF VIOLENCE

Awareness that primates can be distractingly different from our fellow animals—and unnervingly similar to ourselves—has been pervasive throughout civilizations. Carl Sagan wrote of this inclination: "I think the discomfort that some people feel in going to the monkey cages at the zoo is a warning sign."

Like today, in many ancient cultures, monkeys were seen as clever tricksters to keep an eye on, lest they steal your apple and pelt you with a turd. Among the indigenous Akawaio people of South America, there is a mythological monkey figure called Iwarrika known for his selfishness and relentless curiosity. As the folklore goes, Iwarrika is believed to have unintentionally flooded Earth by interfering with the building of a dam by the demigod Sigu. In ancient Japan, monkeys were once considered revered mediators between humans and gods. But at some point, Japan's cultural and religious influencers changed their minds and came to see monkeys as deviants. One ancient idiom implies they are lowly animals trying to be humans but falling short, which, from a biological perspective, is not altogether wrong.

Documented ape encounters are rare throughout history. The few we have are logged with some combination of awe, confusion, and repugnance. Over 2,000 years ago, the Carthaginian explorer Hanno the Navigator encountered what were probably gorillas along the African coast. In his words, they were "savage people, the greater part of whom were women, whose bodies were hairy, and whom our interpreters called Gorillae." Hanno assumed the apes were a more hirsute, wilder ilk of man and tried capturing them. Nabbing the males proved impossible given their strength, but his men did capture and kill three females, sending their pelts home to Carthage.

Around 1590, while imprisoned in equatorial Africa by the Portuguese, English sailor Andrew Battel described two types of man-like "monsters" that would warm themselves by the fire once the local tribes had moved on. "Their faces, hands, and ears are without hair . . . when they walk on the ground it is upright . . . they sleep in trees, and make a covering over their heads to shelter them from the rain." Battel was almost certainly referring to chimpanzees or gorillas, maybe both.

By the mid-nineteenth century, studying and pontificating on our natural history was all the rage in America and Europe. In 1847, an American physician and missionary named—yes—Thomas Staughton Savage and naturalist Jeffries Wyman first described the gorilla in detail to the modern world. They deemed the animal *Troglodytes gorilla*, or, roughly translated, "cave-dwelling hairy people."

By this point, monkeys and apes were being captured en masse—prodded and pondered, studied by scientists and naturalists, enclosed in zoos. In 1863, Darwin's friend and colleague Thomas Huxley observed that humans and gorillas have strikingly similar anatomy and share more characteristics than either species do with monkeys. As Darwin wrote a few years later in *The Descent of Man*, "Our great anatomist and philosopher, Prof. Huxley . . . concludes that man in all parts of his organization differs less from the higher apes, than these do from the lower members of the same group. Consequently there 'is no justification for placing man in a distinct order.'" He goes on to reproach other naturalists of the day for not accepting the idea that humans are a sub-distinction of primates, also recognizing other similarities among us. "Many kinds of monkeys have a strong taste for tea, coffee, and spirituous liquors: they will also, as I have myself seen, smoke tobacco with pleasure."

In the early twentieth century, apes went mainstream, starring in novels, comics, cartoons, and films with a range of demeanors, from a rage-fueled King Kong to the nurturing yet violent fictional species that raised Tarzan. Our opinion of them was a blur between primal violence and untouched innocence. As researchers like Goodall carried out their work, we realized apes are a little bit of both.

Human violence is frequently viewed through the ideas of English philosopher Thomas Hobbes and Genevan philosopher Jean-Jacques Rousseau. Hobbes proposed that the state of nature among presocietal humans was wild anarchy, a "war of all against all." He believed our inner Teen Wolf was tamed only with the formation of organized governments, which he called leviathans. Rousseau, though he didn't coin the phrase, is often credited with the idea of the noble savage, that humans and other apes are inherently good, only to be tainted by the influences of civilization.

Goodall and her colleagues observed the chimpanzees in Gombe National Park for years. Throughout nearly the first decade, the group appeared relatively functional, perpetuating the ape reputation as peaceful. As many researchers have witnessed, the day-to-day life of chimps can resemble a modern-day, slightly chaotic Eden. There's a lot of lazing around. Goofing off. Having sex. All to the tune of frequent and frenetic screeching and laughing. Goodall's initial observations supported the idea that perhaps other apes really do live in a primate utopia, which humans have somehow squandered through greed, technology, and politics.

Then, in 1974, Goodall's field assistant Hilali Matama witnessed something brutal. A group of eight chimpanzees—seven males and one female—quietly crept into a neighboring chimp community where a young and vulnerable male named Godi dined alone in a tree. Godi jumped down and tried to flee. One of the intruders grabbed his leg and dragged him to the ground, pinning the victim like a grappling wrestler. The others proceeded to bite Godi and pummel him with stones. He was left to die, bleeding and screaming as his assailants returned home in victory.

Chimpanzees, especially the males, fight all the time over sexual partners, food, and the most desirable tree limbs and terrain.

Three adult male chimpanzees in a grooming session. Basie, in the middle, is yawning while his older brother, Bartok grooms him.

And killing other members of your species out of one-on-one competition is normal behavior in the animal kingdom. But Matama's encounter was the first time anyone had witnessed any species other than humans come together to coordinate an attack on a member of a neighboring community. This wasn't a random, isolated act. The chimps were at war.

A previously united community known locally as Kasekela had split into two factions. Those that broke off moved to a nearby territory and came to be called the Kahama. Over the next four years, Kasekela males killed all six adult Kahama males. They also killed at least one Kahama female and violently kidnapped another three. The new faction was obliterated.

Much of a chimp's daily routine is leisurely and free of violence. But as many researchers have now observed, every few weeks in some groups, males (and occasionally a few females) will carefully explore the boundaries of their territory. They'll walk in single file, quietly listening for neighbors worth battling. It's called border patrolling, a kind

Chimpanzees border patrolling

of chimpanzee reconnaissance. If while on the prowl, they encounter a lone male or a group smaller than theirs, they'll attack. If evenly matched or outnumbered, they'll give up for the day and return home.

Like Goodall's research group, University of Michigan anthropology professor John Mitani has witnessed how startlingly human chimpanzee violence can be. For years, Mitani has studied a wild chimp population at Uganda's Kibale National Park. Over a period of twenty years, he's witnessed males regularly heading out on these raids, killing over thirty nearby chimpanzees. "The chimps I study here have the dubious distinction of being the world record holders for killing other chimps," he tells me dryly.

But in 2009, something curious happened.

The males among the community he studies began relocating to an area northeast of their usual terrain. What's more, they started bringing females and children along with them. "They were going up there en masse with everyone," says Mitani, "They were yelling and shouting and acting like it was their land."

Mitani realized a lot of the chimps systematically killed over the years had been from this new territory. Those in the group he studies had waited for years until they'd reduced their neighbors' numbers to such a degree that it was safe to move in. It was a land grab. In the end, the invaders increased their acreage by over 20 percent. "They were just up there yesterday enjoying it all," Mitani tells me. "The trees have been producing a lot of fruit lately."

I ask him if the presence of human observers interferes with chimp behavior. He says it does take a while for them to acclimate to the presence of peering researchers, but that for the most part people blend into the woodwork. "They're obviously aware of our presence. They move around us; some like to interact with us and some don't. Mostly it's business as usual," he says. "They'll happily kill an enemy right in front of you, jumping in and out of a pile, biting and stomping them. They can be real crazed lunatics sometimes."

If this sounds a lot like the near-constant boundary disputes we humans have embroiled ourselves in since civilization began, it's because our respective darker sides are biologically entangled, if distantly. Richard Wrangham arrived at Gombe in the early 70s as a graduate student, just as the Kahama and Kasekela were beginning to split. "I didn't witness the war but I witnessed the buildup" he told me. "I saw the first dead body as a result of the conflict."

Wrangham is a leading figure in the study of ape behavior and argues that the aggression seen in chimps is related to the origins of human violence. Given the exceptional nature of our respective brutality toward members of our own species and our highly similar genomes, I think he's right. "That chimpanzees and humans kill members of neighboring groups of their own species is . . . a startling exception to the normal rule for animals," Wrangham writes in his 1996 book *Demonic Males: Apes and the Origins of Human Violence*. He and his coauthor, Dale Peterson, suggest "chimpanzee-like violence preceded and paved the way for human war, making modern humans the dazed survivors of a continuous, five-million-year habit of lethal aggression."

Prior to researchers witnessing the battle at Gombe, war was thought to be uniquely human, a result of civilization and religious differences. As Wrangham put it, "The Kahama killings . . . made credible the idea that our warring tendencies go back into our prehuman past."

WE LIVE IN A SOCIETY

In his 2019 book *Civilized to Death*, author Christopher Ryan argues that modern human civilization is ruining our lives. Many of our advances, he notes, were either developed to address problems we created in the first place, or are used to nefarious ends. We invented vaccines to prevent diseases that were never a problem until we started domesticating animals. We invented airplanes and immediately used them to drop bombs and perpetuate war. Despite the common refrain that, prior to the agricultural revolution 10,000 years ago, early humans were filthy, warring brutes struggling to survive in a cruel world, Ryan says, "There is no evidence to suggest the foraging way of life is less sophisticated or satisfying than any other—including our own—and having lasted hundreds of thousands of years, it's certainly more sustainable." Ryan is open to Rousseauian notions that early hunter-gatherers lived in more egalitarian, more gracious communities, and it wasn't until we bound into civilizations and began farming that large-scale wars and concepts of greed and ownership would come to pervade human existence. As our stuff became more important to us, and as agriculture supported a surge in population growth, we paved over our hunter-gatherer paradise and became, as he puts it, "the only species that lives in zoos of our own design."

Harvard psychologist Steven Pinker takes a more Hobbesian view, famously arguing that humans are now living in the most peaceful era in the history of our species. His interpretation of available data has human violence declining significantly as civilizations coalesced. It's not that we became better people overnight; it's more that the formation of states transferred conflict from the individual to the governing powers. Tribal skirmishes were repressed by armies and law enforcement and replaced by larger-scale atrocities like war, slavery, and genocide. Our violence became centralized and broader in scope.

As humans transitioned from hunter-gatherers to agriculturalists, there was more social exchange, and we wound up living in larger, more permanent groups, cities, and kingdoms. With more interaction and communication may have come more empathy. We recognized shared experiences and struggles, and became more familiar with Others. Pinker believes that alongside the terrible ills society brought—of which surely the worst were the exploitation and enslavement of laborers and treating women as property—concepts like empathy, sympathy, and fairness also trickled through and perhaps even swayed leaders to enact policies that reduced violence.

The rise of commerce and the exchange of resources across longer distances and among more people further solidified social roles and familiarity. If you held an honest job selling goods or services you were appreciated, less expendable to your neighbors, less likely to be a target. Violence wasn't always in our best interest. As Pinker writes, "Daily existence is very different if you always have to worry about being abducted, raped or killed, and it's hard to develop sophisticated arts, learning, or commerce if the institutions that support them are looted and burned as quickly as they are built."

A 2016 study by Spanish ecologist José María Gómez looked at rates of lethal violence in over 1,000 mammalian species, including in over 600 human populations ranging from the Paleolithic era through today. His findings show primates are a violent lot. "A certain level of lethal violence arises owing to our position within the phylogeny of mammals." Phylogenetics is the study of evolutionary relationships among species. In other words, our violence is genetically linked to that of other mammals, especially primates.

Gómez found humans are six times more likely to be lethally violent with each other than most mammals, though about as violent as expected for a primate. Human violence appears to have increased during Paleolithic times and soared during the murderous Middle Ages. In the past few hundred years it then plummeted, supporting Pinker's theory that, despite our wars, mass-shootings, and genocides, we're living in an exceptionally peaceful time compared to our relatively recent ancestors. Gómez found that as humans moved from tribes to chiefdoms to contemporary states our rates of violence dropped dramatically.

The Gómez data show prehistoric bands and tribes had the high levels of violence expected for primates. In some modern foraging communities, however, violence is much higher, possibly because (as Ryan would argue) these populations are becoming more dense as modern culture and deforestation move in. Gómez proposes the increase in violence among early primates was related to increased group living and territoriality as well. They were forced to account for increased population density through conflict.

So was being a lethal killer adaptive and naturally selected for in the great apes? In chimpanzees, probably yes. When male chimps murder neighboring males they're reducing their sexual competition; by taking over more land they're providing themselves, their mates, and their offspring with more resources to support survival and reproduction. Machiavellian rule means selfish genes are more likely to live on. Violent behavior is writ into chimpanzee DNA, and by extension ours. Human morality and self-control are there to rein it in. As Wrangham believes, and as I'll get to, extreme violence wasn't an optimal survival strategy for all apes, us included.

Early life trauma and socioeconomic factors like growing up in an impoverished area are both major contributors to violent tendencies. People who are physically abused before age five are at a higher risk of later-life arrest for both violent and nonviolent crime, not to mention an increased risk of both physical and mental illness.

Genetics may play a role as well. Criminal behavior is much higher in adopted children whose biological parents have a criminal record (Kendler, 2014). Studies show a person's degree of aggressive behavior is under the influence of genes that affect oxytocin, testosterone receptors, and neurotransmitters.

Though the science is young, a few specific genes have been associated with violence. A 2015 study looking at crime in Finland found up to 10 percent of violent crime is committed by individuals with two specific gene variants. The first is a version of the *MAOA* gene, also known as the "warrior gene." It codes for the monoamine oxidase A enzyme, which breaks down neurotransmitters once they've been released. In people with a low-activity version of the *MAOA* gene, dopamine, serotonin, and norepinephrine all accumulate, causing abnormal brain activity, especially in the emotional, fearful amygdala. The other genetic contributor to violence is a version of the *CDH13*—or cadherin 13—gene, known to be associated with substance abuse and ADHD. People with both the *MAOA* and *CDH13* variants are thirteen times more likely to have a history of repeated violent behavior.

Genetic research shows the human genome is still evolving in many ways. Pinker raises the question of whether or not natural selection continues to act on aggression, or, instead, whether changes in how we treat each other are now mostly influenced by culture. He concedes selection might still be acting on aggression but doesn't believe there's enough evidence to prove it. In recent times, cultural pressures seem to have had far more sway in suppressing our violent proclivities than our genomes. He cites rapid changes in societal morality, like the abolition of slavery, the civil rights movement, and an embrace of the LGBTQ community. These were major cultural milestones that took place over just a few generations, far too quickly for our genes to have played much of a role.

Like chimpanzees, humans have an innate predilection for violence. But if touched by civilization for long enough—and the morality and sympathy that can arise out of societal living—rates of violence go down. Pinker writes that part of the pacification process was a reduction in violence toward women. Human men are still dreadfully violent toward female partners, but much less so than male chimpanzees. The World Health Organization estimates that about one in three women worldwide have experienced either physical or sexual intimate partner violence or sexual violence involving someone they don't know. Nearly

100 percent of wild female chimps endure regular violent beatings from males. Even among peaceful bonobos, males frequently attack females, though to less avail given women hold the social power.

Some ape researchers have questioned whether the violence we see in today's wild chimpanzees is a result of increasing encounters with humans. Across Africa, logging and other industries are decimating chimp habitats in the interest of profit. Maybe the apes are growing angrier and more violent as we encroach on their land. Maybe they're just pissed off.

Nope.

The research shows chimps are just really violent sometimes.

In 2014, a team of thirty researchers analyzed 152 killings among eighteen chimpanzee communities with human contact. The vast majority of the killings were perpetrated by males against other males. Over 60 percent of the attacks were against those from other communities. Human interactions had a negligible influence on the murders compared to adaptive strategies. Offing male neighbors improves a chimp's chances of success and survival (Wilson).

Chimpanzee violence seems innate and unavoidable. In our species, we have a brain more able to sequester our apish demons.

CHAPTER EIGHT

OUR SOFTER SIDE

The bonobo reputation swings opposite that of chimps. These are our more matriarchal, more agreeable cousins. A reflection of why humans can be good people.

The species wasn't even known to science until well into the twentieth century. Anatomically, they more or less look like small chimps and were originally known as "pygmy chimpanzees." In the early 1920s, psychologist and early primatologist Robert Yerkes bought what he thought were two chimpanzees named Chim and Panzee from a zoo. Both lived with him at his home in New Hampshire where they dined at a small table and ate with forks. Yerkes noticed distinct differences in their personalities, writing in his 1925 book *Almost Human*, "In all my experience as a student of animal behavior I have never met an animal the equal of Prince Chim in . . . alertness, adaptability, and agreeableness of disposition." Yerkes noticed Chim was especially sensitive, reasonable, and intelligent, referring to him as an anthropoid genius. It's now assumed Chim was actually a bonobo, kinder and gentler in his ways than Panzee.

Bonobos were first recognized in 1928 by a German anatomist, and confirmed as their own species a few years later by American anatomy scholar Harold Coolidge, who, after examining a fossil collection from a museum in Belgium, noted their more delicate features and smaller adult skull. It then took decades for anthropologists to really appreciate

Chim (left) and Panzee with their original owner, Mr. Noel E. Lewis

the distinction between chimps and bonobos. "When I got into this business in the late 60s, people still didn't fully realize there was a difference between them," recalls Tattersall. "There were a few in captivity in zoological parks, but it wasn't until fieldwork started that the distinction was more widely understood."

The first researcher to extensively study bonobos in the wild was Japanese primatologist Takayoshi Kano. Kano set up camp near the village of Wamba, now in the Democratic Republic of the Congo. The local villagers were friendly and he could hear bonobos calling in the nearby forest. So he planted some sugarcane.

Over the course of many months, the apes gradually crept out of the forest and Kano was able to observe their ways. He noticed it was adult females holding court over the sugarcane patch, not the expected alpha males, who call the shots in most mammal communities. Kano was struck by how leisurely and peaceful the Wamba bonobos were— lots of grooming, snacking, lazing around. Occasionally he observed a male get angry, but the dominant females would either ignore him or gang up to chase him away.

In the decades since, field observations by Kano and other primatologists have given us a good sense of the bonobo persona and social structure, but it was Frans de Waal who really cemented their reputation as the gentler ape. De Waal directs the Yerkes National Primate Research Center in Atlanta and believes bonobos can express compassion and kindness. Compared with chimpanzees, they're more sensitive, and, in their willingness to share food, altruistic. Male chimps will fight with each other over estrus females (those "in heat"), while male bonobos seldom do so. Male chimps frequently commit infanticide, killing unrelated babies in hopes of impregnating the mom themselves. Not the case among bonobos. Nor is violent gang warfare against neighbors common. Male bonobo conflicts can be plenty violent, but far less so than those observed in chimps, and often involving some degree of peaceful denouement.

Female chimpanzees and bonobos differ in their behavior too. Whereas female chimps tend to stick to themselves, especially when not in heat, female bonobos prefer to aggregate in groups. They will also set out to explore new territory, something not often seen in chimp communities.

It was Kano who first noticed the bonobo preoccupation with sex, especially as a means of calming tensions and male aggression. Bonobos are the only animals besides humans known to tongue kiss and have sex facing each other. Many of them, especially the younger ones, engage in an orgy of sexual acts in which anything goes. Men with women. Men with men. Women with women. Old and young. I won't go into detail, but one popular activity is called "penis fencing." Sex between male and female bonobos is in part driven by females spending much more time in estrus than chimp females. This means they're more often sexually receptive. With increased sexual opportunity, competition and aggressive behavior among males goes down. Mitani likens the morning routine of wild chimpanzees to a rowdy frat party. "They yell, fight, and bang things." Bonobos, rather, rise and before long are probably getting it on in one way or another; at very least, they're not roaming off to murder their neighbors.

De Waal proposes our shared evolutionary history with bonobos as a means of understanding our own morality. He believes our sense of right and wrong results from day-to-day social interaction acting on our innate biology, arguing that our moral values are not some divine construction, but have been evolutionarily ingrained in us since the dawn of our species. The genomes and neural circuitry of social primates have evolved to accommodate group living. As we go through life interacting with others, our moral code is subject to revisions based on cultural information.

This theory plays out on our smartphones daily as cable news and social media rewrite our moral and cultural guidelines faster than ever before. What's right versus wrong can change on a dime. "Moral law is not imposed from above or derived from well-reasoned principles; rather it arises from ingrained values," De Waal writes. To the best of our knowledge bonobos aren't religious, yet they exhibit moral behavior all the time.

De Waal believes our natural history isn't entirely a narrative of male dominance and xenophobia, but also one of "harmony and sensitivity to others." He feels attributing human progress, if we can call it that, to aggressive men having won more battles against other aggressive men—which was hugely influential in shaping our past—is short-sighted. "Attention to the female side of the story would not hurt, nor would attention to sex." What if we didn't necessarily always conquer our human enemies, but in some cases joined them? Our shared DNA with Neanderthals and Denisovans proves that, to a small degree, this did happen. "I wouldn't be surprised if we carry other hominin genes as well. Viewed in this light, the bonobo way doesn't seem so alien."

Brian Hare got the bonobo bug when he took de Waal's course as an undergraduate at Emory. In the 2000s, he wound up as Wrangham's graduate student, collaborating with his mentor at Lola ya Bonobo, a bonobo sanctuary in the Democratic Republic of the Congo. Wrangham and Hare conducted some of the first research directly comparing chimpanzee and bonobo behavior. When they first started, Hare had already worked with chimps for over ten years and went into his first bonobo encounter confident. The social dynamic wasn't what he'd expected.

"It was totally different. The females were very skeptical. They wanted nothing to do with me!" he recalls.

"If you play with male chimps or bonobos, and to a lesser degree female chimps, they become engaged. They want to be around you," says Hare. "You're their buddy." Not so with female bonobos. Hare's now-wife Vanessa, a science journalist, had tagged along to the Congo for the chance just to witness apes in the wild. When Hare had trouble connecting with the females, plans changed. "She ended up having to do all the science for us since we couldn't do it ourselves," he jokes. "The females loved her!"

It's not as if a man can't work with female bonobos, it just takes more time and patience. A study by Wrangham and Hare looked at grooming and playing behaviors among adult male and female bonobos and chimpanzees. Most of the apes preferred to interact with someone they had previously spent time with—except for female bonobos, with whom playing or interacting did little for the long-term relationship.

By comparing the bonobos at Lola ya Bonobo to previously studied chimps from the Ngamba Island Chimpanzee Sanctuary in Uganda, Hare confirmed that bonobos are more congenial, less selfish, and more likely to cooperate with each other. When presented with food, both chimps and bonobos are equally good at cooperating to obtain it. They'll join up to pull a tray of chow closer to their cage. However, chimps will only do so with a degree of civility if the meal is divided into shareable portions. The moment the food is lumped into a pile that can be monopolized, everything goes to hell—Hare describes the scene when separate piles of ape treats were combined as "total chaos." Bonobos presented with the same situation cooperated just fine.

Wrangham believes the differences in behavior may be due to chimps evolving in regions scarce in food, whereas bonobos developed in the lush, nutrient-rich Congo basin. Perhaps chimps had to evolve extreme selfishness in order to survive. Bonobos, in their "giant salad bowl," as Hare puts it, had plenty of food and free time.

When Kano originally arrived in Wamba, local villagers regaled him with bonobo folklore (Kappeler, 2012). The story goes that there was a younger brother in a family of bonobos who grew tired of forest life and raw food. Roaming the rainforest in tears, he met a spirit who taught him to make fire. He moved onto the plains and began cooking his food, his descendants becoming humans. His older brother stuck to the traditional ape lifestyle and his descendants remained bonobos. Despite Africa's active bushmeat trade, killing or eating a bonobo is traditionally taboo in Wamba. We're family, after all.

There are only so many ape sites left to be studied in the world; and only so many researchers with the resources to get to the jungles of Africa and Southeast Asia to study them. Every new observation is valuable and can reveal some previously unknown morsel of ape character. And anthropologist Martin Surbeck has made a few.

Surbeck directs the Kokolopori Bonobo Research Project in the DRC. Started in 2016, in collaboration with local Congolese residents and the Bonobo Conservation Initiative, the Kokolopori project involves monitoring two bonobo communities to better understand their physiology and behavior.

Bonobos have a mama's boy reputation, which has garnered them scorn from macho anthropologists in the past. Whereas bonobo daughters migrate to other communities and regions, sons stick by their mothers through adulthood. In 2019, Surbeck published a study showing that bonobo moms guide the sex lives of their sons. In covering the research in the *Atlantic*, science writer Ed Yong recounted one of Surbeck's more memorable bonobo encounters: "Two of them—Uma, a female, and Apollo, a young, low-ranking male—were trying to have sex. Camillo, the highest-ranked male in the group, caught wind of their liaison and tried to come between them. But Hanna, Apollo's mother, rushed in and furiously chased Camillo away, allowing her son and his mate to copulate in peace."

Surbeck used DNA analysis to determine who was related to whom among four wild bonobo and six wild chimp communities. He realized bonobo mothers were acting in ways that influenced whom their sons mated with. While many mammal mothers will help guide the reproductive success of their daughters, that bonobo moms do so with their sons supports the idea that females play a strong role in their society. If a son stays close to his mom, he becomes part of an influential social circle. "It's not that the moms are actively picking their sons' mates. It's more that the powerful moms in charge of the group get the best spots. They're first in the trees, at the center of things. If a son sticks by his mother he'll have a better chance at copulation," says Surbeck.

There's yet another twist to the socio-sexual drama that is bonobo life. In one bonobo community Surbeck observed, 60 percent of the offspring were sired by one male. It's almost as if the dominant females had designated—or at least supported—an alpha male, yet one who would never outrank the alpha women. The anointed lad was sexually successful but would always play a subservient role in the community. Bonobo social roles channel and allow for the sexual voracity of young males, but in a way that minimizes violence against females and supports stability.

"Here's how I explain it to myself," Surbeck told me. "There are some females who are very high ranking—they're at the center of their community. I think there is an element of other females choosing to mate with the son of a high-ranking female. But no matter how successful he is at copulating he won't rise above his mother or the other powerful females."

Much like Hare, Surbeck believes bonobos' sex-crazed reputation is exaggerated, or at least misinterpreted. In both chimpanzees and bonobos—just like in humans—there are certain individuals who have a lot of sex and others who don't. Females from outside groups will often use copulation to gain acceptance into a new community. But there are also high-ranking mothers who may not have any sex at all. Some degree of promiscuity in both species can be strategic. Among chimps in particular, moms have to look out for infanticidal males. By mating with more than one partner, she clouds the paternity, giving multiple males the idea that they might be the father. The same polyamory occurs in bonobos, with sexual confusion diffusing male competition and shifting power to the females.

Given our culture's history of toxic masculinity and the macho brutality seen in chimps, I asked Surbeck if we'd be better off in a female-driven social system more akin to bonobo society. "Oh, this becomes a very personal question," he hesitated. "I don't think I can answer this based on what I'm studying. But I think it's naive to think things would be that much better." He points out that when in power, bonobo females adopt some of the same stereotyped behaviors we associate with powerful men. Aggression. Bullying. Coercing.

"Maybe to a certain degree everything would've been better with females as the leaders, but then you see these bonobo mothers intervening in an almost male-like manner pushing for their sons' reproductive benefits." Surbeck says that while female bonobos restrain male violence, primatologists don't actually know if they are any less violent than female chimpanzees. Among apes, perhaps power itself, regardless of sex, is the corrupter of pacifism.

The idea of bonobos as peace-loving flower children and chimps as crazed warriors is reductive. The observational and comparative evidence suggests as much. Both male and female bonobos can be plenty violent when defending their territory—biting, fighting, and throwing sticks when necessary. As he gained field experience, Surbeck was more surprised at just how cooperative chimpanzees can be given their unpolished reputation. When not murdering each other, male chimps are quick to support their peers and are actually more likely to have "friends" and form social groups than male bonobos. Yes, these can be the same coalitions that gang up to raid and ravage neighboring chimp communities. But they also look out for each other, much in the way human friend groups do. These are the same blocs humans have formed since forever ago.

"How much they support each other was so surprising to me," says Surbeck. "If someone is injured in the chimp community, everyone cares, including the males—who we think of as so violent. And they are. They still fight and kill. But they also band together and reconcile more so than bonobos do." According to de Waal, chimps will even reconcile conflict with a hug and a kiss. Despite their more peaceful reputation, bonobos generally don't form these tight social bonds. They're less likely to depend on their community members for support and seem to care for others a little bit less. I like to think of them as an echo of the loners among us.

De Waal is resolute that some of our kinder qualities are shared with both chimpanzees and bonobos. He writes of female chimpanzees dragging stubborn males toward each other to make amends after a scuffle, taking away any weapons in the process, and of male chimps arbitrating conflicts within the group. "I take these hints of *community concern* as a sign that the building blocks of morality are older than humanity." And further, that chimps, despite being incredibly violent at times, have a softer side too.

We share the vast majority of our DNA with both chimps and bonobos, and our best and worst social tendencies seem tangled up in our shared genomes. It's impossible to say whether we got this quality or that directly from a shared ancestor with either species, but we're offered a glimpse into aspects of our personalities by observing theirs. "Cognitively speaking, I wouldn't say it's as simple as us being some combination of chimpanzees and bonobos, but we share most of our genomes and as a result, it would be surprising if many bonobo and chimpanzee behaviors didn't exist in us in some way, at our best and worst moments," says Hare.

Maybe we can learn a little more about conflict and community warfare by studying chimps; and more about egalitarianism and peace by studying bonobos. "As it goes when you're trying to find differences, you often find yourself swimming in a sea of similarities," Hare adds.

Wrangham has written of the seemingly nice aspects of blatantly terrible people, "We have a rare and perplexing combination of moral tendencies." Hitler was said to be agreeable and friendly, Pol Pot a soft-spoken and kind educator, Stalin calm and gentlemanly. "We can be the nastiest of species and also the nicest."

CHAPTER NINE

SPEAKER WIRE

When a chimpanzee stomps her feet it often means "I want to play." A loud screech translates as "I want to be groomed." A tap on your shoulder? "Stop it, I don't like what you're doing."

Not only do chimps use various hoots, grunts, and shrieks to communicate, they also employ well over eighty physical gestures with at least nineteen different meanings. These can all be found in the *Great Ape Dictionary*, an ever-growing online compendium of chimp and bonobo gestures run by primatologist Catherine Hobaiter. Between 2007 and 2009, Hobaiter spent eighteen months observing chimpanzee communication in Uganda's Budongo Forest Reserve. Since then, she and her team have continued to study ape communication, adding new gestures to the dictionary as their meanings are confirmed. Hobaiter thinks there is a clear link between ape communication and the evolutionary origins of human language.

Some chimpanzee motions are intuitively familiar to us. Chimps will fling their hand at someone to say "go away" or reach out a palm to request food. Others look like human gestures but have different meanings. A hand in the air isn't a call for attention, but rather an ask for another chimp to move closer. And some have completely different meanings than we're accustomed to. "There is something about spinning around that seems to mean 'stop that,'" says Hobaiter. "Whether they do a pirouette or a somersault, it's all the same." She admits such a meaning is hard to confirm definitively since, in the case of one chimp suggesting another changes their behavior, many gestures seem to work. Think of the human equivalent: an eye roll, a head shake, grabbing someone's arm. All of these can convey we're displeased with what someone else is doing.

Hobaiter and University of York psychologist Kirsty Graham found the vocabulary and meaning of gestures between chimps and bonobos overlap almost perfectly. The only major differences between the species are the frequency of social and sexual gestures. "Yes, there's a lot more use of requests for sex in bonobos! But these requests don't differ much from the same gestures used by chimpanzees," says Hobaiter. They're just used more often.

Beginning in the 1960s and continuing through the 1980s, it was faddish to teach apes to communicate like humans. Using sign language and lexigrams (symbols representing words), orangutans, gorillas, chimpanzees, and bonobos were all able to learn somewhat extensive vocabularies. Washoe the chimpanzee, Kanzi the bonobo, and Koko the gorilla were among those who mastered hundreds of signs and symbols. Critics of ape language research are unconvinced this means much of anything. Many feel that learning a lot of words falls short of using language, and the field has moved toward studying the gestures and sounds apes use on their own, rather than those we insist they learn.

By observing how apes communicate in the wild, scientists believe we can better understand our own social history. Among the suite of social traits and behaviors that influenced hominin brain evolution, language and symbolism may have been the most influential.

"To me language is an obvious stimulus," says Tattersall.

"It was probably our ability to appreciate symbolism, including language," Holloway adds.

Dean Falk is especially convinced. "In terms of our brain evolution, I come down hard that the thing that really drove it was language; it was so important in getting the dominos set up for hominids going where they eventually went."

Noam Chomsky held a brutally reductionist view of how human communication evolved into verbal fluency. The famed linguist was known for his theory that a single random mutation occurred 50,000 years ago and gifted Sapiens with language. He argued that since no other species is capable of syntax, the ordering of words to form sentences, there is no evolutionary precedent for language in other animals. But by the time of his 2016 book *Why Only Us*, Chomsky's stance had softened, writing with his coauthor, computer scientist Robert C. Berwick, that language may have appeared closer to 200,000 years ago. Unlike earlier in his career, he also now believes Neanderthals might have had language. The discovery of a Neanderthal hyoid bone from an Israeli cave suggests they too could speak. In humans, the hyoid supports the base of our tongue and helps enable speech. More so than in other primates, the fossil shows the Neanderthal hyoid was positioned as it is in Sapiens. They may have been capable of some form of spoken word (D'Anastasio, 2013).

MIT linguistics professor Shigeru Miyagawa speculates the evolution of language traces not just to primate communication, but also to that of chirping birds. Like birdsong melodies, our voices can modulate to expand on a finite vocabulary. Darwin was of like mind, writing in *Descent of Man*, "The sounds uttered by birds offer in several respects the nearest analogy to language."

Chomsky argues we are born with innate language ability, which arose once in one species. Steven Pinker agrees we have a universal capacity for language, but believes its evolution followed a more gradual Darwinian course, driven by certain key mutations along the way, like those allowing for syntax. There are four laws of behavioral genetics that together describe how genes influence our behavior. The first three are:

- All human behavioral traits are heritable (affected by genes).

- The effect of being raised in the same family is smaller than the effect of genes.

- A substantial portion of the variation in complex human behavioral traits isn't accounted for by the effects of genes or families.

In short, they say human behaviors are heritable, but that depending on our upbringing, the environment also shapes the differences between us. If two brothers grow up in the same household—or society or type of environment—their genes play a big role in any differences between them. If, for whatever reason, they grow up in two different households, their surroundings—their nurture, not their nature—have much more authority over their personality and behavior (Turkheimer, 2000). The fourth law was proposed by a team of behavioral geneticists, which includes Pinker's former student James Lee. It states that most heritable behavioral traits are a result of many genes working together, each with only a very small effect on its own.

"Though a single gene can disrupt a psychological trait, no single gene can install one," says Pinker. "This is consistent with the mechanism of natural selection, in which a beneficial ability, because it is statistically rare, is astronomically unlikely to have arisen by a single lucky mutation." He believes the same explanation goes for language acquisition.

Early humans used an array of gestures, utterances, and Paleolithic grunts that had taken on meaning—call them proto-languages, akin to babies sputtering out "ma-ma" and "da-da." Selection would've favored increasingly verbose communication, as the ability to convey information could only have benefitted human survival. "Though it doesn't add up to language as we know it, sophisticated communication involving gesture, body language, and sound must have been there in early humans," says Tattersall. He believes the common ancestor we shared with Neanderthals and Denisovans was preadapted to eventually evolve symbolic speech.

Preadaptations, or exaptations, are traits that evolve for one purpose but are then co-opted for another. It was Darwin who first floated the idea that the function of a trait might change through generations. More recently, the late paleontologist and biologist Stephen Jay Gould popularized the idea; his favorite example was bird feathers. Birds had feathers long before they used them to fly. Initially, they were used to attract mates and provide insulation. Early birds were preadapted for flying before they actually took to the skies.

Similarly, Tattersall and others believe our brains were language-ready long before we spoke. As Wayne State University linguistics professor Ljiljana Progovac wrote in critiquing Chomsky, "It is entirely possible that the initial (simpler) stages of language recruited elements of existing genetic makeup." If so, the evolution of language called on genes and abilities already in place, resulting in the exaptation of meaningful syntax and speech. Selection, Progovac feels, may have acted on the genes of individuals who were "just a little better at combining words or storing words in the memory."

In 2002, researchers reported on a gene called *FOXP2* that appeared to be Chomsky's language gene. In a study of thirty-one family members, fifteen were afflicted with a severe speech disorder. All fifteen carried a mutant version of *FOXP2*. A subsequent paper reported humans carry two variants of the gene and both must be functional for normal language to develop. These variants aren't found in other primates, implying they arose and spread quickly through the human population within the last 200,000 years. The researchers concluded *FOXP2* must've played an important role in human language acquisition (Enard, 2002).

And it probably did. But it's far from the only genetic influence. The original study has never been replicated and only included genetic data from a small group of individuals, most of whom were of European and Asian descent. By looking at a wider array of genomes, including those from people of African heritage, researchers found that harboring the *FOXP2* variants wasn't necessary for language (Atkinson, 2018). Since then several other gene variants have been associated with language acquisition, some present in Neanderthals and Denisovans and others not (Mozzi, 2016).

Also supporting the gradual evolution of language is monkey anatomy. In 2016, scientists reported the vocal tracts of Macaques are speech-ready and could theoretically produce comprehensible sounds if the monkeys only had the neural ability to speak (Fitch). For decades it was assumed monkeys aren't able to make human-like speech due to their vocal anatomy. Using X-ray videos of macaques communicating, Fitch and company found their vocal tracts have an architecture amenable to producing the broad range of sounds necessary for speech. This, they believe, implies changes in neural organization—changes within the language centers of the brain and their connections to neurons that control the muscles of speech—may have been more important to the evolution of human speech than previously thought. The vocal structure revealed by X-ray is one capable of producing five distinguishable vowel sounds, the same number that most modern human languages use. With a slight modification in neurocircuitry, macaques would be capable of uttering their a's, e's, i's, o's, and u's (though considering what they'd have to say to us after all these years of monkey experimenting, maybe it's best they can't).

Also supporting the speech capabilities of monkeys is the infamous mirror neuron system. The original mirror neuron research by Rizzolatti confirmed mirror neuron activity in a macaque brain region called F5, a smallish patch of neurons toward the front of the monkey's head that correlates with a human language center called Broca's area. The thinking goes that if F5 mirror neurons enable the mimicry of meaningful gestures, maybe they primed the brain for the evolution of vocal communication. In monkeys, the F5 region seems to help produce and read facial expressions, just as our language circuitry does. Broca's area and a brain region called Wernicke's area are essential partners in how we process speech and language. The former sits toward the rear of our left frontal lobe, behind the temple; the latter is farther back in our temporal lobe, also on the brain's left side. Broca's area helps produce speech, while Wernicke's helps us comprehend it. In humans, communicating through symbolic language is a neurological dance, one that involves many other regions of the brain as well—those influencing personality, emotion, and the control of facial muscles. But the fact that two important regions are only present on one side of the brain highlights the importance of asymmetry.

The left and right hemispheres of our brain are not mirror images of each other. A host of structures and functions are unique or predominant on one side or the other.

And it's unclear why. It may be that compartmentalizing certain functions to one side makes neural transmission more efficient. Information doesn't have to travel as far as in a bilateral setup. The pop-science adage that the left brain controls analytical thought and the right brain, creativity, has been largely discredited. Neuroimaging and autopsy studies on artists and mathematicians show little difference between hemispheric brain structure or activity. Van Gogh's right brain wouldn't have been any more active than Einstein's. Visionary painting and theories of relativity require interconnected and asymmetric regions from all over the brain.

Monkeys and apes have many of the same asymmetries as we do, but to a lesser degree, suggesting lateralization was influential in human brain evolution. Studies show many primate asymmetries are heritable, coded for by our genes. Yet in humans, the genetic influence is less than in chimpanzees. Environmental influences have more say our in our development; our brains are more plastic in response to the deluge of sensory information we're exposed to through life, including the strange sounds our parents make while we're infants that coalesce into language (Gómez-Robles, 2016).

Monkeys and apes both have the equivalent of Broca's and Wernicke's areas, but our increased brain plasticity allows us to better absorb and reproduce the symbolic sounds and vocal patterns we hear. Our speech centers are also more complex at the microscopic level. Semendeferi and other researchers have found that compared with monkey language centers, the human equivalent has more horizontal space between neuronal columns, leaving more room for connectivity and more sophisticated language production and processing. Recent research confirms that among primates, humans have the highest degree of neuronal connectivity in Broca's area, followed by other apes and, to a lesser degree, monkeys (Schenker, 2008; Palomero-Gallagher, 2019).

The social brain hypothesis depends on communication among group members. Let's assume Dunbar's idea that human's increased social intelligence and complexity can support a larger group size is to some degree correct. He believes this is due in large part to language and sophisticated vocal gestures, which he calls vocal grooming. Through plain old physical grooming, chimps can maintain around fifty friends, family members, and tolerable acquaintances. With language, that number jumps to around 150, where it remains today in humans. Being able to convey "I think I might like you," or "We're cool, right?" with a few symbolic sounds, rather than having to delouse your friends for much of the day, frees up a good deal of time for other pursuits.

Dunbar believes that, since social group size and complexity correlate with brain size, we reached his number of 150 sometime between 2 million years ago and the appearance of *Homo sapiens* 200,000 to 300,000 years ago. In monkeys and apes, there's a direct relationship between group size and time spent socially grooming. As our social groups grew along with our expanding brains, there would've been a glass ceiling effect in which there wasn't enough time in the day to stabilize a growing network of relationships through grooming. Dunbar's data show that as hominin social circles grew, the time needed to maintain bonds through grooming would've exceeded the twelve hours of available daylight on the African savanna by over 50 percent. Without another means of socializing we couldn't have increased our group size beyond that of other primates.

Our social lives became a lot more interesting—and influential to the fate of the planet—once we could speak to each other. There was a new and more efficient way to communicate. Information, both social and ecological, could spread faster than ever before. Dunbar believes more complex vocalizations were the last in a line of preadaptations for speech, the others being laughter and singing. He argues both laughing and song tap into this same endorphin system and helped grow our social groups, bonding us tighter through shared experience.

A study from 2017 on which Dunbar was a coauthor, supports his hypothesis (Manninen). The researchers used an imaging technique called PET, or positron emission tomography, to measure laughter-induced release of endorphins in adults watching comedy movies with their friends. Social laughter boosted opioid release in a number of brain regions, particularly the cingulate and orbitofrontal cortices, both involved in emotional responses and social intelligence. They also found watching comedies as opposed to dramas with their friends was associated with a higher ability to tolerate pain. They believe modulating our opioid signaling system with a good laugh may help reinforce relationships.

With language came the inevitable: gossip. Many researchers over the past few decades, Dunbar included, believe gossip was hugely important to hominin evolution. It's thought the tendency to jabber behind one another's backs helped maintain stability among early hunter-gatherer populations and may still help keep the peace within communities today. When humans talk, over 60 percent of the time we talk either about ourselves or other people: what they're doing, what they're wearing, why Dave from finance is so weird. Gossip serves as a community watchman. It keeps tabs on the cheaters and conmen and boosts the reputation of the more honorable among us and those who contribute to the group. Passing judgment on those around us may seem inherently rotten, but it served an important purpose (Dunbar, 1993).

British philosopher Julian Baggini calls gossip a "moral appraisal of other people . . . it's about the judgment of what people are doing and whether it's right or wrong, good or bad." In *The Descent of Man,* Darwin wrote, "After the power of language had been acquired, and the wishes of the community could be expressed, the common opinion how each member ought to act for the public good, would naturally become in a paramount degree the guide to action."

E. O. Wilson might say bonding through laughter, language, and gossip took hold around the fire, which humans first harnessed as early as 1.5 million years ago. As we settled into campsites, late-night banter may have helped solidify the community. A toasty flame emanating light extended the day into the evenings, allowing for more social interaction after a day on the plains hunting and gathering. A 2014 study by anthropologist Polly Wiessner analyzed the day and nighttime conversations of the Ju/'hoan (!Kung) bushmen of northeast Namibia and northwest Botswana. During the day, most conversations were practical, involving topics like land rights and economics. At night, more than 80 percent of their conversations were stories, dealing with culture, people, and other communities. Fire, language, and social bonding may have all gone hand in hand as we stayed up late sharing tales around the embers. Throw in some dancing, singing, and religious rituals, and you have the social and emotional glue of community.

We don't know for sure when language arose, but neuroscience tells us the neurocircuitry of complex communication exists in primitive forms in the brains of other primates. Chimps have their gestures and make up to 100 meaningful sounds in the wild. Tattersall feels Australopiths probably had a slightly larger vocabulary but didn't have much appreciation of symbolism. The skull of *Homo habilis*, the first known human species, has more prominent ridges overlaying Broca's and Wernicke's area, implying the two critical language centers were enlarged. Brain size and a large Broca region suggest *Homo erectus* also probably communicated using a form of protolanguage. Some researchers believe that if members of *H. erectus* were indeed seafarers, as it appears they were, they would've needed a sophisticated language to source materials, build boats, and navigate open waters.

Pinker points to the ideas of John Tooby and Leda Cosmides, who direct the Center for Evolutionary Psychology at the University of California, Santa Barbara. Tooby has said, "I believe human evolution consists of exploration of the 'cognitive niche' in which the unusual problem-solving abilities, language, and sociality of *Homo sapiens* coevolved, each multiplying the benefits of the other two."

Tattersall believes humans acquired modern symbolic language around 100,000 years ago, right before we got really creative. Prior to this, ingenuity was rare. Symbolism grew out of an increasingly social existence driven by communication. Our genes, neurocircuitry, and anatomy had prepped us—preadapted us—for the spoken word and culture to come.

CHAPTER TEN

TEMPERAMENT OF THE DOG

When a mouse stands petrified in the gaze of an eagle, its amygdala is on fire.

This ancient paleomammalian brain center is no more than two almond-shaped blebs buried deep behind the temples. Yet, for nearly 200 million years, it has been one of the most important agents of survival in mammals. The amygdala processes emotion, especially fear and anxiety. At its core, fear in the animal kingdom is pretty straightforward: either flee and hopefully don't get eaten, or stick around and fight. The amygdala is the air traffic controller behind the fight or flight response. In humans, it fires not just out of mortal fear, but also when we're anxious. We've evolved social and cultural "dangers" that other animals don't have to deal with, like public speaking, work stress, and online dating.

Everything our eyes, ears, and nose detect in our environment is relayed directly to the amygdala for evaluation. If incoming sensory information suggests danger, it signals to our hypothalamus, which triggers a hormonal cascade causing the adrenal glands to release cortisol, our "stress hormone," as well as epinephrine (adrenaline). This is why we often feel energized and hyperaware during fearful situations—mental clarity ups our chances of getting through danger. We're stressed, yet alert.

Stress also activates the sympathetic nervous system, another key player in whether we fight or flee. When we're scared, our sympathetic system causes our pupils to dilate and our muscles to tighten in case things come to conflict. At the same time, vegetative functions like digestion, controlled by the parasympathetic nervous system, are momentarily put on hold, to be resumed once we've avoided the saber-toothed cat. Or once we've made it through the speech.

The amygdala helps etch sensory memories into our brains for life. It's the reason why, in a split second, certain sights, songs, and smells can trigger emotions and recollections all the way back to childhood.

Long considered abstract God-given qualities, emotions were first proposed to have a biological basis by Darwin. In his 1872 work *The Expression of the Emotions in Man and Animals*, he wrote, "Certain actions, which we recognise as expressive of certain states of mind, are the direct result of the constitution of the nervous system." The father of evolution believed some feelings had a physical component that stemmed from the structure of the brain; he also believed that, like most other traits, our emotions underwent natural selection and were shaped by evolution.

His ideas hold up today. Not only are specific brain regions and circuits known to process and draw up certain mental states but, as Darwin also recognized, emotions often come with corresponding facial expressions—a link between physicality and feeling—and animals exhibit primitive parallels to many of our emotions, suggesting an evolutionary trajectory. Though we like to think of profound emotions as primarily human, other animals are far from dead inside. Birds obsess. Elephants and chimps may grieve. And that thing when your dog woefully watches you pull out of the driveway from the window—that might be DSM-certified separation anxiety.

Emotions like fear, along with sex drive, are among the most essential and primitive mental states. One promotes survival and the other procreation, the two pillars of natural selection. Motherly affection for offspring, which scientists believe arose in later mammals, also makes sense from a survival standpoint. By the time hominins were gossiping around the campfire, our social lives were likely taking on the much more profound emotions we feel today. Our emotional selves evolved hand in hand with social intelligence and communication. Whatever it means to be human, it surely involves the beautiful and tangled mess of feelings we all feel. As social primates evolved increasingly complex brains, our capacity for deeper feeling also increased. Higher cortical regions expanded and intensified our emotions; new brain centers connected with old ones to bring fresh dimensions to our primitive emotional pangs. We wound up with grief, guilt, envy, pride, and love.

Now when we're scared, our prefrontal cortex is called in to negotiate with the amygdala. "I know I'm terrified, but what's the worst that can happen?"

"Imagine them all not wearing pants."

Our cortical reasoning ability gives context to our fear as we approach the podium. Anxieties and carnal feelings now manifest in step with higher cognitive interpretation. As with language, there was a course of evolution to our emotions, which we can trace in other species. Still, as far as we know, humans have the most highly developed array of emotions on Earth. We are emotional wrecks, and presumably this benefits us in some way.

In the 1970s, professor Paul Ekman conducted research backing Darwin's ideas. His work confirmed the universality of facial expressions across cultures. It also suggested that prior to taking on communicative meaning, certain mugs were simple environmental adaptations. When afraid or startled, our eyes open wide to expand our field of vision. The wrinkled nose of disgust helps prevent noxious toxins from being inhaled. Only later, did these reactions become exaggerated to communicate surprise and repulsion. According to psychologist Carroll Izard, the development of language was a big player in emotional evolution. We could now share our experiences and feelings and better plan to collectively maximize positive emotional experiences in the future (returning to a plentiful fruit tree), while avoiding potentially negative experiences (not returning to the lion's den you'd stumbled across yesterday). Emotion, entwined with symbolic language and gossip, helped maintain and gauge social relationships. As we banded into groups to, say, forage, a moocher who felt guilt and learned his lesson might get another chance at friendship and survival. One who brazenly hogged all the plums without remorse might not—and would most certainly be a major topic of discussion around camp later that night.

It's incredibly difficult to study the biology of something as complex, confusing, and marvelous as why we feel the way we do. Much emotional research has compared emotions that clearly exist in both humans and other animals to parse our similarities and differences. Not to take anything away from the vast body of research on the more primitive emotional states, but fear and sex drive at their most fundamental level are the low hanging fruit of emotional research. Whereas a feeling like love is an immensely more complicated psychic journey—an abyss of biology and psychological theory that many might prefer science stays away from. If we explain away the beauty and mystery of humanity's most beautiful, intangible qualities with a bunch of neurons and chemicals, the thrill is gone.

But let's try anyway.

As ape and hominin babies were being born earlier and earlier due to the obstetrical dilemma of having a big brain, stronger bonds were necessary to keep helpless children alive. Selection favored genes for bonding between not just mother and child, but family, community, and mother and father. The extreme of human bonding may be our willingness to die to protect our friends, community, or country. There are few patriots in nature. The human capacity for tight social bonds is unparalleled, and I'd like to believe it's both a product of a genetic tendency toward altruism—even if caused by our selfish genes—and the more recent cultural influences that gave us a sense of morality.

The neurobiology of why we feel emotions like lust and love is byzantine. Lust is, of course, more primal, a drive to reproduce and pass on our genes driven by our hypothalamus signaling to the testes and ovaries to produce testosterone and estrogen; testosterone is the main driver of sexual desire in both men and women, but estrogen also helps increase female libido, especially during ovulation.

During sex, our brain's reward system works in tandem with our sex hormones to produce pleasure, but it also serves as a kind of bridge between pure lust and more profound attachment. When we experience pleasure, both the hypothalamus and a primitive region of the brainstem called the tegmentum release dopamine to other regions of the brain, including the striatum, the nucleus accumbens, and the amygdala. Together the network reinforces attachments and bonds; it draws up those feelings of euphoria that come with being in love. In humans, the recently evolved prefrontal cortex also gets pulled into our sense of reward, giving our bonds more depth than other apes presumably feel; we're capable of intense romantic bonds and of deeper emotional pain when a romance falls apart. Research shows that both love and lust act on our reward circuitry, yet in slightly different ways. Sexual desire activates an area of the striatum associated with pleasurable vices, like drugs and food. Loving feelings activate a different region of the striatum plus part of the cortex called the insula, which helps give a pleasurable experience meaning. Love arises from a higher, more abstract cognitive appreciation of pleasure and bonding (Cacioppo, 2012).

Working with the reward system to strengthen close relationships we also have oxytocin. This "cuddle hormone" not only strengthens the bond between mother and child—causing the nursing reflex and the release of milk—it's also released during intimate physical contact of any kind. Whether sexual, romantic, or platonic, it creates in us a warm social euphoria. Mice who've had the expression of their oxytocin gene blocked no longer recognize other mice they once knew. Give them a shot of the hormone and their social lives return to normal almost immediately (Winslow, 2002).

GOOD GRIEF

The counter to love and attachment is grief, our response to loss. Some see mourning as an evolutionary side effect of attachment. When we lose someone or something so important to us—whether a relationship built on romantic love, social enjoyment, or an altruistic commitment to community or cause—how can we not feel deeply shattered?

Love isn't necessarily an addiction, but it shares the tendency toward obsession and compulsion. When we lose someone, it can look a lot like withdrawal from an addictive drug. The pain, emotional and physical, is real. In a letter to his mourning cousin, Darwin wrote, "Strong affections have always appeared to me, the most noble part of a man's character and the absence of them an irreparable failure; you ought to console yourself with thinking that your grief is the necessary price for having been born with such feelings."

After I'd grown up and moved out of my parents' house, they lived for a few years next to a cattle farm in Barboursville, Virginia. I still remember being home for various holidays and hearing the plaintive, guttural cry of the mother cows when their calves were taken away and sold to another farm. There's no telling what the moms were thinking, but I'm convinced they were feeling some form of sorrow and mourning. They missed their babies.

Chimpanzees have been observed demonstrating something close to grief, which may inform the biology behind human mourning. There's the well-known instance of a Zambian chimp community's reaction to the death of a nine-year-old group member, taken too soon by a respiratory infection. Researchers studying the group had named the young male Thomas. Within minutes of his death, twenty-two members of his forty-chimp community had convened around the body, sniffing and touching his corpse. They were eerily calm and quiet for chimpanzees, sitting and contemplating the death, even when presented with food. At one point an adult female ran over and slapped the body, as if asking, "Is he really dead?" or maybe telling, "Look, he's dead! Let's get on with things." An adult male called Pan, who'd cared for Thomas for years as an adoptive father, was distraught, frantically running around screeching and guarding the body from the others. A chimp named Noel moved in and cleaned Thomas's teeth with a blade of grass.

It's hard to say for sure if the reaction to Thomas's passing is definitely a form of grief, or if the chimps in his group were, for certain, mourning. But chimps are known to take far more interest in the remains of their friends and families than those of other members of their community.

Barbara King believes many animals demonstrate grief, and that a large brain size and lots of cognitive power is not necessary to rue loss. Elephants show intense interest in their dead, with members of different families traveling to the bodies of deceased elders, engaging in a sniffing and touching ritual similar to that of Thomas's community. The well-reported saga of the orca J35 has also been interpreted as a mother in mourning. The whale carried the corpse of her dead calf for seventeen days and 1,000 miles before finally releasing it. Marine biologists were convinced this was a mom unwilling to let go of her deceased newborn. King has written of grief occurring in species ranging from ferrets to donkeys to elephants, and she's especially concerned with the grief reactions experienced by farm animals, whose offspring are routinely taken away from them at a young age (like those cows living next to my parents).

Swiss-American psychiatrist Elisabeth Kübler-Ross famously divided the grief process into five stages: denial, anger, bargaining, depression, and acceptance. Some version of this emotional journey comes with an evolutionary rationale. When we grieve, we feel actual pain and discomfort. We're scared and anxious, and our cortisol soars in sorrow and stress. It's only natural to feel anguish when a biology favoring strong emotional bonds is dealt such a blow, but if we're going to survive and be mentally capable enough to maybe care for more children in the future, we must eventually move on. We accept. Some psychologists argue grief is the sign of someone capable of emotional commitment and depth, a desirable partner. Others believe it's a teaching mechanism—the pain you feel when your child falls off the ledge and drowns means you won't leave another child unattended near that ledge ever again.

Similar to grief, many complex emotions probably served an adaptive function. University of Texas at Austin evolutionary psychologist David Buss proposes feelings like guilt, envy, schadenfreude, and gratitude probably served some purpose to the individual and community (Al-Shawaf, 2016). These emotions may have solved adaptive problems like determining social and sexual hierarchies, morality, punishment of freeloaders, and protection of one's family. Our emotions, and the problems they solve, are more complicated than a basic, "I'm scared, so I'm running away."

DOMESTIC BLISS

The importance of emotions in driving our social evolution is equaled by our ability to restrain them. Despite the rampant cruelty and brutality humans exhibit daily, we are shockingly good at controlling our temperaments compared to most other animals. There's a popular idea circulating through anthropology, recently propagated by Richard Wrangham, that humans have the ability to keep our cool because we are a domesticated species, much like dogs. Whereas dogs were intentionally domesticated by humans, we underwent a process of self-domestication through evolutionary adaptation that, in our better moments, allowed us to control our aggression. In *The Goodness Paradox*, Wrangham breaks down our aggressive behavior into two distinct types, "reactive aggression," violence in response to provocation, and "proactive aggression," or violence that's deliberately planned, like premeditated murder and war. Chimps exhibit both types of violence regularly. But in humans, reactive violence has plummeted compared with apes and earlier hominins. Wrangham uses the example of an airplane—pack a 747 with chimps and there's no outcome but violent chaos. Thousands of humans fly with each other peacefully every day—yet we are still forced to go through an invasive security screen before boarding, since some rogue Sapien might be plotting to blow up the plane. We are the most skilled species on Earth at enacting premeditated, devastating acts of violence—yet compared to our wild ape counterparts, we're angels.

Wrangham writes that domestication doesn't equate with tameness. People have tamed wolves, only to have them turn around and, for lack of a better word, go *ape*. The same goes for chimpanzees. Plenty of people have lived with them over the years, most famously Michael Jackson, who shared his Neverland Ranch with Bubbles the chimp. But even the tamest of chimpanzees are far from domesticated.

No story of chimp cohabitation gone wrong is more notorious than that of Travis. A local celebrity in his Connecticut town, the erudite thirteen-year-old male chimp could water plants, use a television remote, and unlock doors. One day Travis left the house with his owner Sandra's car keys and she couldn't convince him to come back inside. Sandra's friend Charla came over to help, bringing an Elmo doll, one of Travis's favorite toys. The scene that followed was horrific. Charla lost her hands, face, sight, and part of her brain to the attack. When the cops arrived, Travis walked up to the police car, opened the driver-side door, and was shot by the officer inside. He made it back to Sandra's house and died next to his bed.

This isn't the behavior of a domesticated animal. Domesticates are genetically distinct from their wild ancestors and are much better at dealing with others, including strangers. Dogs will occasionally attack humans, but often as a result of behaviors instilled or encouraged by their owners and seldom with the ferocity that wolves (or chimps) are capable of. Work by Michael Tomasello and Brian Hare shows that, though chimps are far smarter, dogs are much more attuned to human social cues (Kaminski, 2009; Hare, 2002). And puppies who've had very little human contact are better at understanding human signals than full-grown wolves raised from birth by humans are. After 15,000 years of breeding dogs to live by our side, they're good at listening to us.

"I was blown away when we discovered that dogs could cooperate and communicate with humans in ways that chimps cannot," recalls Hare. "Chimps are brilliant at so many things, and can solve problems that dogs could never solve. But this really basic ability to understand the intent of a gesture is really difficult for them."

The idea that humans are somehow more domesticated than other animals isn't new. It was considered by Aristotle, early anthropologist Johann Friedrich Blumenbach, and Darwin. Darwin realized domesticated animals are not only more docile than their wild counterparts, like dogs versus wolves, they also have a set of common but unrelated traits. Domesticated animals often have smaller bodies, floppy ears, and curlier tails. Their faces tend to be recessed, with smaller jaws and teeth than wild animals, and differences between the sexes are less pronounced, with males becoming less masculine and the sexes physically converging.

Modern research has shown wild animals also tend to have larger amygdalae, which may explain their more fearful, more aggressive reactions (Kruska, 2014). Overall the brains of domesticated animals tend to be smaller. Based on anatomical differences among hominins, Wrangham believes humans underwent the domestication process and became more docile around the time *Homo sapiens* arose 200,000–300,000 years ago. This could explain why our brains have actually been shrinking for the last 10,000 to 30,000 years. We came down with "domestication syndrome." Wrangham feels that as humans formed increasingly cooperative societies, evolution would have favored slightly less aggressive, less physical males—there was selection against the bullies. A premium on being friendlier led to a more agreeable, less emotionally reactive human species.

In his 2020 book, *Survival of the Friendliest*, Hare elaborates. "The idea of 'survival of the fittest' as it exists in the popular imagination can make for a terrible survival strategy. Research shows that being the biggest, strongest, meanest animal can set you up for a lifetime of stress," he writes, adding that constant aggression comes with costs, like upping your chances of being killed.

In the late 1950s, Soviet geneticist Dmitry Belyaev began studying silver foxes in Siberia to better understand the process of domestication. Fox fur was valuable around the world, and Siberian families had kept them for generations. Belyaev began selectively breeding the more docile foxes, those less fearful of humans. Within just three generations, the pups were noticeably less scared and aggressive around people. By the fourth generation they would run up to humans wagging their tails like dogs. In subsequent generations, Belyaev noticed many fox families had developed a white, or piebald, spotting pattern on their fur. White spots set to a colored background is a classic mark of domestication—think

Dmitry Belyaev with his foxes

of the black-and-white patterns seen on cows, dogs, horses, and white-footed cats wearing "socks." Eventually, Belyaev's foxes developed floppy ears, and the males' skulls shrank to be less aggressive looking and more like those of females. Years later, while working with Wrangham, Hare traveled to Siberia to study the storied foxes. He found that though they'd never been bred for their cognitive abilities, the experimental foxes were showing dog-like skills in reading human social cues. It seems selecting for docility and domestication brings with it a host evolutionary byproduct traits that don't appear to have any necessary function, much like the male nipple.

Some researchers believe domestication may have helped pave the way for language. Domesticated bird species tend to have lower levels of corticosterone, their primary stress hormone, analogous to our cortisol. In birds, high levels of the hormone are associated with impaired cognition and slowed growth of the song-learning system. In some bird species, selection for docility and friendliness is associated with chirping out more complex songs. If a mellower environment allows for more sophisticated sound production in domesticated finches, maybe a lower-stress, domesticated environment did the same for humans.

Neural crest cells are another interesting piece of the domestication puzzle. These are a temporary strip of cells running down the backs of vertebrate embryos. During development, the cells migrate to help form the peripheral nervous system, the cartilage and bone of our face and skull, and the melanocytes that give our skin pigment. They also migrate to form the adrenal glands. Selection for domestication and docility reduces adrenal size and hormone production by reducing neural crest migration, which in turn reduces aggression and emotional reactivity. By selecting for stunted neural crest migration, the gamut of domesticated traits comes along for the ride. This explains all those nonadaptive traits that seemingly do nothing but accompany domestication. With smaller populations of neural crest cells reaching the outer limits of the developing body, the teeth, jaw, and cranium are smaller. Ears get floppy as structural cartilage doesn't extend quite as far. Domesticates are friendlier, more delicate creatures, to the point where you might even equate the domestication syndrome with "juvenilization." Both emotionally and physically, they have more childlike traits. This perhaps explains why Sapiens resemble less-intimidating Neanderthals. Wrangham believes bonobos underwent this process, accounting for their more agreeable temperament, whereas chimps did not. "I don't know what brain changes occur between a six-year-old chimp and a sixteen-year-old chimp, but my guess is that, in terms of anatomy and temperament, bonobos are much more like the six-year-old chimp," he says. Wrangham isn't convinced we owe our peaceful qualities directly to our shared ancestry with bonobos, but instead that our similarities were a case of convergent evolution, the concept of the same trait evolving independently in multiple species. Being less reactive may have been beneficial to both of us docile apes and therefore selected for independently in more than one species.

The migration of neural crest cells is in part driven by a gene called *BAZ1B*. Most of us have two copies, yet people with Williams-Beuren syndrome only have one. Those with this rare genetic condition have cognitive impairment, small skulls, elf-like facial features, and are extremely friendly; a strikingly similar set of features to the domestication syndrome. In a study published in late 2019, researchers looked at the effects of increasing or decreasing *BAZ1B* activity in a line of neural crest stem cells. Tinkering with the gene influenced hundreds of other genes known to be involved in facial and cranial development, and reduced *BAZ1B* activity was found to be important in driving the facial features seen in Williams-Beuren syndrome. The researchers also looked at hundreds of craniofacial genes in DNA from modern humans, Neanderthals, and one Denisovan. They found the human genome had undergone extensive mutations to help regulate the genes' functions. Many of these same gene variants have also been under selection in domesticated animals, supporting the theory that humans underwent a recent domestication (Testa).

In 2018, Kent State University anthropologists C. Owen Lovejoy and Mary Ann Raghanti reported that we humans have a unique cocktail of neurotransmitters in our brains that may also have contributed to our self-domestication. The team measured neurotransmitter levels in brain samples from humans, chimpanzees, gorillas, baboons, and monkeys, all of whom died of natural causes. Specifically, they tested levels in the striatum, which you'll recall is involved in social bonds, romantic behaviors, and reward. While humans, gorillas, and chimps all showed elevated striatal serotonin activity, compared with the other species, humans had markedly increased striatal dopamine. Elevated striatal levels of both neurotransmitters are associated with increased cognitive and social intelligence. Also in the mix is the neurotransmitter acetylcholine, higher levels of which are associated with aggression. Lovejoy and Raghanti found gorillas and chimps have much higher levels of acetylcholine than do humans. "The high striatal serotonin shared by humans and great apes likely contributes to the cognitive flexibility required for complex social interactions," Raghanti says. "The lower acetylcholine in humans corresponds to our decreased aggression, compared to most other apes. It's a concert really."

Raghanti and Lovejoy believe the human brain's neurochemical profile was shaped by natural selection due to the various reproductive and survival benefits it conferred. The team speculates that our elevated striatal dopamine and serotonin levels would have led to more advanced social behaviors, encouraging selection for empathy and language. Before the hominin brain began its rapid enlargement two million years ago, our canine teeth had already shrunk, from fangs to the modest points we have today. This indicates we'd become less aggressive, consistent with the neurotransmitters found in our brain. Our self-domestication resulted in part from a unique neurochemistry compatible with social understanding and restrained emotion.

With higher cognition came a puzzling tug-of-war between emotional and rational thought. Humans excel at making bad decisions even when we know they are wrong. Truth can matter little in matters of family, friends, and politics. We support our factions unconditionally. But why? Did this paradoxical thinking benefit us evolutionarily? Why do we listen to our gut or to our heart?

You could argue that certain guttural emotions were adaptive. Sure, snapping at someone in anger makes you a jerk and can have negative consequences. Or it can help retrieve a stolen mammoth steak as you aggressively grab it back and secure some calories. And perhaps it's worth believing in falsehoods and sun gods if it ingratiates you into a supportive, protective community.

It's hard to shake millions of years of rash emotional reactions. Our instincts pull from brain centers both primitive and novel; our decisions are divided between animal reactivity and human reason. Despite having far more self-control than most species, our primitive brain often wins. Our social and emotional lives are at once unique and shared, as our loves, losses, embarrassments, and angers are reflected in the psyche of other animals—our relationships and symbolic language are a distant elaboration of a grooming Macaque or a stomping chimpanzee.

Our social lives even grow old like those of other primates (Almeling, 2016). Like us, as monkeys age, they become more socially selective, preferring the companionship of their friends to those less familiar or those they consider social bores. In their dotage they'd rather spend time with old pals than deal with the small talk—small grooming?—required to go out and make new friends. Beloved crank and comedian Larry David once told an interviewer he tolerates people like he tolerates lactose—which is to say, I'm assuming, not well. Like many elder humans, monkeys often become cranky old folks, set in their ways, their daily routine, and those individuals they can tolerate.

PART THREE

FOOD, FIRE, AND THE FUTURE OF THE HUMAN BRAIN

It was a strange world ahead that would unfold,
a thunderhead of a world with jagged lightning edges.

—BOB DYLAN

CHAPTER ELEVEN

WEATHER PERMITTING

Humans will eat just about anything, within reason.

We are among the most omnivorous species on Earth, and on at least two occasions this has saved our butts.

As Earth entered the Quaternary period (the geologic period we're still in) 2.6 million years ago, the planet began a cycle of periodic warming and cooling. Every 40,000 years, and later every 100,000, glaciers would creep down from Greenland and cover much of North America and Europe. Then they'd gradually retreat. By analyzing ancient flora found in sediment, scientists have shown that, in Africa, the waffling climate led to drastic shifts in vegetation and landscape. During cooler, drier stretches, the lush forest gave way to woodlands and vast savannas. One cool spell toward the beginning of the Quaternary period coincides with the extinction of *Australopithecus afarensis*, Lucy's people, and the appearance of our genus, *Homo*. The tall, big-brained *Homo erectus* showed up 600,000 years later. Scientists believe the new species were those who could not only best the competition, which would've been very important, but also endure the changing climate and adopt a varied diet. They adapted to the new terrain, increasingly traveling longer distances on two legs to forage, scavenge, and eventually hunt. In *Scientific American*, Columbia University paleoclimatologist Peter B. deMenocal wrote, "The creatures that adapted to these shifts—those that showed flexibility in what they ate and where they lived—appear to be the ones that prospered."

Increased social intelligence would've helped hominins endure climate change and improve caloric returns. Being social provided safety in numbers in our newly exposed home on the plains. "The social angle is exquisitely important," says Barbara King. "After all, hunting *is* a social activity, as is gathering and probably scavenging too. I believe a key factor in our evolution was an increasing selection pressure for cooperation." The social brain became a necessity.

Our new lifestyle also favored those who could run—Australopithecines gradually went from being short, stocky, and hunched to lean and leggy. By the time *Homo erectus* came along, hominins were adeptly bipedal and covering more distance than ever before, exposed to new adaptive pressures and turning to new food sources. African terrain went through many fluctuations between woodlands and plains, but the generally drier, wide-open environment led to the migration of *erectus* out of Africa and into Asia as early as 2 million years ago.

Our food supply has always thrived or failed at the mercy of Mother Nature. Radioactive dating of rocks and fossilized teeth support the theory that at least two drastic shifts in climate forced us to either expand our palate or perish. And some hominins didn't make it. Dental fossils show that during these periods of climate change, our ancestors moved away from a primarily vegetarian diet to an omnivorous diet with a balance of meat, fruits, and vegetables. Our *Paranthropus* cousins—the robust Australopithecines—were not as lucky. As climate change rolled in, they appear to have remained vegetarian. With forest fruits scarce, they turned to largely indigestible grasses. By about one million years ago, they were gone.

Evolutionary endurance requires a proper diet. Using fossil findings and radioactive dating, scientists from a variety of fields now have a pretty good idea of what our ancestors ate and of which foods and nutrients may have played a role in building our brain. Our unique and expanding diet—and our evolving ability to access foods other creatures couldn't—is a key part of our story.

Monkeys live in a three-dimensional arboreal world where their big eyes and keen sense of their surroundings help them find calories in fruits, seeds, and flowers. They are incredibly skilled at this, even using their little fingers and opposable thumbs to pluck out individual seeds from a picked-over fruit. For them, selection for intelligence may have involved both social pressure and the ability to acquire food in trees. Monkeys are also ecologically plastic, catering their diet to changing environments, and are happy supplementing fruit with insects, fungi, and small animals when they have to.

Apes who evolved in tropical regions are also dependent on the fruits of the canopy, supplementing their diet when times are scarce. They love a good bug-covered stick. And they've been known to use crude tools to dig up edible roots, the discovery of which is typically a big event resulting in lots of screeching and competition for the prize. Chimpanzees will also kill the occasional monkey for its meat. The males form premeditated hunting parties in which one chimp chases a monkey out of a tree while the others wait at the bottom to corner it. Bonobos sometimes engage in the same strategy, but with more female involvement.

"Many primates and apes live in tropical habitats and it takes a certain amount of intelligence to know where all the food is," says John Mitani. He says chimps are very aware of the state of their food supply. "I know their forest very well and each and every year think it's going to be this fruit tree or that area over there that will produce most of the food," he laughs. "But it always ends up being somewhere else. It takes smarts to adapt to these seasonal and yearly changes. They have incredible knowledge of the sources of food around them."

The earliest hominin diet was probably very similar to that of chimpanzees. For a time, our early ape ancestors would've gone back and forth between savanna and woodland, finding calories wherever they could. If a drought dried up the plains they could head to the trees and dine on fruits, seeds, and monkeys; if the forest was over-picked, it was back to the savanna to dig for roots. Fibrous, nutrient-rich bulbs, tubers, and corms were a steady, reliable source of nutrition for early humans; they grew nearly everywhere and were protected from less savvy animals by growing underground.

THE MEAT OF IT

Compared to other apes, Sapiens have very small teeth and a reduced jaw. It's thought this change occurred in part to accommodate for language, with a smaller instrument being better capable of controlling sound. Our mouths and molars also evolved as a result of our changing diet. In early hominins, *Orrorin* through *Ardipithecus* to *Australopithecus*, and even more so in the genus *Homo*, our canines—or "fangs"— shrank from being large and ape-like to more Sapien-like as we no longer relied on our mouths as a weapon. Australopithecines ate soft fruits like apes do, yet had molars capable of processing nuts, seeds, and roots. They were more omnivorous than many other apes, but maintained only a thin layer of protective dental enamel like chimpanzees. This may explain the sorry state of most fossilized Australopith teeth, not yet fully adapted to the tougher fare of the savanna. In general, as we move through hominin evolution, the jaw and teeth shrink in size, while tooth enamel thickens to handle a more diverse diet. Through *Homo habilis* and *Homo erectus* we lose our lethal fangs and end up with a small, compact mouth, crowded with enameled teeth that can process a wide array of foods, from soft and fleshy to hard and fibrous. Our shrunken jaw was a harbinger of self-domestication. As Darwin put it, "The early . . . forefathers of man were . . . probably furnished with great canine teeth; but as they gradually acquired the habit of using stones, clubs, or other weapons, for fighting with their enemies or rivals, they would use their jaws and teeth less and less. In this case, the jaws, together with the teeth, would become reduced in size."

If our preadaptations for communication and social complexity primed our brain to take over the world, it was one soft, fleshy food in particular that provided the fuel: meat.

Meat makes up just 3 percent of the chimpanzee diet, as it probably did for early humans. But as our ancestors gradually moved from forests to fields, we adopted more consistent carnivorism. And in an evolutionary instant, our brains ballooned in size. Bathed in proteins, fats and an abundance of calories, the primate brain grew like never before, culminating in the massive noggins we carry around today. It's impossible to say whether increased meat consumption directly caused our increase in brain size, or whether other adaptions that left us with larger, more intelligent brains meant we became better hunters. Perhaps it was a little bit of both, with access to meat supporting the caloric requirements of evolving a bigger brain.

Barbara King eats mostly vegetarian, yet cedes the importance of steak to our past. "I believe we need to use our evolved brains to think compassionately about who we eat these days. But I still underscore the fact that meat, specifically, played a key role in the evolution of our brains. We can't politically—as vegans, vegetarians, or reducetarians—rewrite our evolutionary history. These are the facts." Even just a small amount of meat is packed with vitamins and nutrients that are healthy for the brain and body. B vitamins, iron, zinc, and selenium are just a few.

In the beginning, increasing our meat consumption meant scavenging. Early humans were not at the top of the food chain like we are today. Before advanced weaponry, we were pretty much helpless against a hungry cat of even modest disposition. What benefited us was how sloppily large predators took their meals (and still do). When predators like lions or wolves take down prey, they typically go straight for the organ meat. The kidneys, liver, spleen, and brain are incredibly dense with nutrients, proteins, and fats—a much more efficient source of nutrition than gnawing on raw muscle. Like those mysterious people who leave half the meat on their Buffalo wings, lions also tend to leave plenty behind. We moved in and nabbed their leftovers, likely fighting off ancestral hyenas looking to do the same.

Anthropologists don't know exactly how carnivorous humans were at first. Modern hunter-gatherer diets are all over the omnivorous map, some high in meat consumption, others not. Fossil evidence—including the Taung Child—suggests hominins as far back as *Australopithecus* may have dabbled with the crude butchering. By the time of *Homo habilis* and later *Homo erectus*, we were most certainly regular meat eaters. In 2013, Smithsonian anthropologist Briana Pobiner and her colleagues documented over 3,700 animal fossils and nearly 3,000 stone tools from Kenya dating back over 1.5 million years. The bones clearly indicate hominins, most likely *erectus*, were consistently breaking down animal carcasses. By this point, meat had become integral to our Paleo food pyramid, and we remain the only primate on the planet with a largely carnivorous diet.

A couple years later, Pobiner traveled to a private game reserve in Kenya to study the potential for early hominins to scavenge as a viable means of life. The preserve was dominated by lions, whose ancestors would've lived alongside *Homo erectus*. Pobiner spent months documenting the eating habits of lions by analyzing felled carcasses, nearly half of which had some meat left on them—in many cases, lots of it. Scavenging what lions and saber-toothed cats discarded would've been a viable survival strategy for us less-lethal humans.

Also on our menu was bone marrow. Fossils show early humans used hammerstones to crack open bones and scrape out calorie-rich marrow, which big cats also tend to leave behind (how ironic that one of our first forays into carnivorism is now an *au courant* delicacy in trendy dining).

At some point, we became less dependent on scraps. We were getting access to the prime cuts. We began competing for meat with other carnivores, probably banding together to scare them off or fight it out. We also began hunting, actively attacking small animals with crude stone flakes and projectiles. By comparing the kill preferences of modern lions with those of early humans, University of Wisconsin anthropology professor Henry Bunn found that humans could stalk, ambush, and kill larger animals as early as two million years ago. Perhaps, he argues, we weren't entirely dependent on scavenging. Prior to his claim, the earliest clear evidence of human hunting was a 400,000-year-old site in Germany where we impaled horses using long spears.

FIRE ON THE PLAINS

The brain is energetically very costly. In the 1990s, Leslie Aiello and Peter Wheeler came up with the expensive tissue hypothesis, proposing that the evolution of our big brains must have been offset by the shrinking of other organs, namely the gastrointestinal tracts and liver. It was around the dawn of *Homo erectus* some two million years ago that our gut started shrinking. We couldn't afford to chew grass and leaves all day or we'd waste away. Meat provided a higher-quality, higher-calorie meal. Aiello and Wheeler proposed it was the muscle of other animals that allowed the human brain to expand and evolve the way it did.

Meat was important. But so was how we prepared it. Richard Wrangham believes it was our harnessing of fire that really drove our brain expansion, writing in his 2009 book, *Catching Fire: How Cooking Made Us Human*, "I believe the transformative moment that gave rise to the genus *Homo*, one of the great transitions in the history of life, stemmed from the control of fire and the advent of cooked meals."

Our earliest experience with cooked morsels would've been reaping the remains of brush fires. The charred muscle and sinew of the smaller animals that succumbed to the flames would've been an easy, tasty source of calories; probably one our ancestors had been enjoying to a modest degree since our split from other apes. Chimpanzees will sometimes pick over brushfire remains in search of cooked seeds. Research by Wrangham and his colleagues shows great apes prefer cooked food, suggesting Paleolithic hominins probably did too (Wobber, 2008). In a series of studies conducted in 2015 by University of Michigan evolutionary biologist Alexandra Rosati and her Harvard psychologist colleague Felix Warneken, captive chimps consistently preferred their vegetables cooked. Rosati and Warneken presented the apes with slices of raw potato, which they then dropped into a bowl. After shaking the bowl, they'd remove a previously planted piece of cooked potato. The chimps quickly associated shaking food in a bowl with a preparation they preferred. Now, if handed a cooked potato, they would gladly eat it. If given a raw slice, they'd drop it in the bowl. If no bowl was available they'd hold on to their slice or move to another area looking for one. They'd developed an interest in cooking their food out of taste preference, as our human ancestors did.

Preserved ash from Wonderwerk Cave in South Africa has humans cooking food at least one million years ago (Berna, 2012). Cooked food is not only easier to chew it's also easier to digest, saving us energy. Putting food to flame destroys toxins and harmful microbes; in certain foods, it releases vitamins and nutrients unavailable in their raw form. The drastic change in our diet would've had serious physical, neurological, and social consequences. Our teeth and chewing muscles continued to shrink as our food got softer. Anthropologist Peter Lucas estimates molars can be up to 82 percent smaller when needed for chewing cooked versus raw potatoes. Cooking gave us a much easier means of accessing nutrients.

We now know diet is closely linked to our brain health and function. Early humans with reliable, nutritive meals may have benefited from a cognitive boost. Improved cognition would've led to more creative pursuits, unveiling new skill sets, advantageous to survival. Populations with better cooks—and better masters of fire—could outcompete those gnawing on raw, chewy gristle. Because cooking helps preserve meat and can render tough vegetables palatable, a fresh kill could go a lot further, and we could partake in a wider array of plants. It's a Darwinian chicken or egg scenario: cooked food provided us with abundant nutrients to support a growing brain, and in turn, our big brains allowed us to cook and create in even more interesting ways. "The discovery of fire, probably the greatest ever made by man, excepting language, dates from before the dawn of history," wrote Darwin. "He has discovered the art of making fire, by which hard and stringy roots can be rendered digestible, and poisonous roots or herbs innocuous."

Cooking food can also, in some cases, reduce available nutrition, denaturing certain nutrients or creating indigestible proteins. But overwhelming evidence shows cooked meat and vegetables are a much more efficient source of energy; and as Wrangham says, "life is mostly concerned with energy."

"So you're not a proponent of the raw food movement?" I asked him.

"You can do it now that we have all of these wonderful, highly domesticated foods available year-round from different parts of the world," he said. "But it takes some work. Raw foodists thrive only in rich modern environments where they depend on eating exceptionally high-quality food." A raw diet often requires enough money to regularly browse Whole Foods and enough time and effort to plan meals rich enough in calories and nutrients. Think of spinach. When sautéed a big leafy bundle cooks down to a fraction of its original size. One cup of the cooked stuff has six times as many calories and a much higher contraction of nutrients than a cup of raw. It also requires a lot less chewing to get down. Purely raw diets have been associated with lower levels of vitamin B-12 and HDL (our good cholesterol), as well as lower bone mass and impaired ovulation in women.

The healthiest diets probably include a mix of both cooked and raw foods. Cooking vegetables like carrots, peppers, and spinach releases antioxidants, like phenols, lutein, and beta-carotene. On the other hand, vitamin C and certain B vitamins are lost in many cooked vegetables. For brain and body health, it's best to mix it up.

Tattersall isn't all in on Wrangham's commitment to the idea of self-domestication, but he appreciates the importance of fire to our past, saying: "He makes a powerful case for cooking in the early processing of foodstuffs to support the expanding brain. The brain is a very energetically expensive organ that would've needed a high-quality diet. One way of getting a high-quality diet is to build fires and cook." Wrangham believes *Homo erectus* was among the first hominins committed to the use of fire. "The reason I say committed is that they had a relatively small gut, the size of ours rather than those of chimps. And their teeth were also small like those of *sapiens*, meaning they had access to relatively high-quality food even during periods of the year when it was difficult to find any food."

As it does today, cooking would've helped early humans preserve food and extend the spoils of a hunt or forage. Wrangham recalls the nine months he spent with a population of Pygmy hunter-gatherers in the Congo. "They killed an elephant during the first few weeks I was there and hung strips of the meat up in the huts where there was fire. They were still eating it when I left."

Once humans had fire, it became nearly as essential to our existence as food, water, and shelter. Prior to the harnessing of fire, humans were much more beholden to nature. When the sun went down, the day was over. Imagine camping with no light or heat source. Nights were dark and dangerous. Everyone turned in early hoping for a predator-free rest. Fire allowed us to stay up late bantering and storytelling, igniting our social lives. It provided protection by scaring away animals; it allowed us to literally smoke small prey out of their holes and to roam into new, chillier geographies in search of undiscovered food and resources.

Cooking may have come with a heavy cultural cost. Many anthropologists believe that, prior to our mastery of fire, women, like female bonobos, enjoyed some degree of clout and dominance within groups. As fire brought more reliable, longer-lasting nutrition, men could spend more time out hunting and honing other skills, while females stayed behind to forage and cook. This new setup was a major disruption to social and cognitive evolution. As Barbara King told me, "Cooking essentially opened up these whole nutrient packages to us, and at the same time, altered social dynamics." The obstetrical dilemma already tied young moms to camp; fire and cooking may have further driven the sexual division of labor still widespread in society today. Cooking, as Wrangham puts it in *Catching Fire,* "trapped women into a newly subservient role enforced by male-dominated culture . . . and perpetuated a novel system of male cultural superiority. . . . It is not a pretty picture."

Men got to roam, with women stuck at home.

CHAPTER TWELVE

A PALEOLITHIC RAW BAR

"He was a bold man that first ate an oyster."

—JONATHAN SWIFT

That man or, just as likely, that woman may have done so out of necessity. It was either eat this glistening, gray blob of briny goo or perish. Beginning 190,000 years ago, a glacial age we identify today as Marine Isotope Stage 6, or MIS6, had set in, cooling and drying out much of the planet. There was widespread drought, leaving the African plains a harsher, more barren substrate for survival—an arena of competition, desperation, and starvation for many species, including ours. Some estimates have the Sapien population dipping to just a few hundred people during MIS6. Like other apes today, we were an endangered species. But through some nexus of intelligence, ecological exploitation, and luck, we managed. Anthropologists argue over what part of Africa would've been hospitable enough to rescue *Homo sapiens* from Darwinian oblivion. Arizona State University archeologist Curtis Marean believes the continent's southern shore is a good candidate.

For two decades, Marean has overseen excavations at a site called Pinnacle Point on the South African coast. The region has over 9,000 plant species, including the world's most diverse population of geophytes, plants with underground energy-storage organs like bulbs, tubers, and rhizomes. These subterranean stores are rich in calories and carbohydrates, and, by virtue of being buried, are protected from most other species (save the occasional tool-wielding chimpanzee). They are also adapted to cold climates and, when cooked, easily digested. All in all, a coup for hunter-gatherers.

The other enticement at Pinnacle Point could be found with a few easy steps toward the sea. Mollusks. Geological samples from MIS6 show South Africa's shores were packed with mussels, oysters, clams, and a variety of sea snails. We almost certainly turned to them for nutrition.

Marean's research suggests that, sometime around 160,000 years ago, at least one group of Sapiens began supplementing their terrestrial diet by exploiting the region's rich shellfish beds. This is the oldest evidence to date of humans consistently feasting on seafood—easy, predictable, immobile calories. No hunting required. As inland Africa dried up, learning to shuck mussels and oysters was a key adaptation to coastal living, one that supported our later migration out of the continent. Marean believes the change in behavior was possible thanks to our already keen brains, which supported an ability to track tides, especially spring tides. Spring tides occur twice a month with each new and full moon and result in the greatest difference between high and low tidewaters. The people of Pinnacle Point learned to exploit this cycle. "By tracking tides, we would have had easy, reliable access to high-quality proteins and fats from shellfish every two weeks as the ocean receded," he says. "Whereas you can't rely on land animals to always be in the same place at the same time." Work by Jan De Vynck, a professor at Nelson Mandela Metropolitan University in South Africa, supports this idea, showing that foraging shellfish beds under optimal tidal conditions can yield a staggering 3,500 calories per hour!

"I don't know if we owe our existence to seafood, but it was certainly important for the population Curtis studies. That place is full of mussels," says Tattersall. "And I like the idea that during a population bottleneck we got creative and learned how to focus on marine resources." Innovations, Tattersall explains, typically occur in small, fixed populations.

Large populations have too much genetic inertia to support radical innovation; the status quo is enough to survive. "If you're looking for evolutionary innovation, you have to look at smaller groups."

MIS6 wasn't the only near-extinction in our past. During the Pleistocene epoch, roughly 2.5 million to 12,000 years ago, humans tended to maintain a small population, hovering around a million and later growing to maybe eight million at most. Periodically, our numbers dipped as climate shifts, natural disasters, and food shortages brought us dangerously close to extinction. Modern humans are descended from the hearty survivors of these bottlenecks. One especially dire stretch occurred around one million years ago. Our effective population (the number of breeding individuals) shriveled to around 18,000, smaller than that of other apes at the time. Worse, our genetic diversity—the insurance policy on evolutionary success and the ability to adapt—plummeted (Huff, 2010). A similar near extinction may have occurred around 75,000 years ago, the result of a massive volcanic eruption in Sumatra. Our smarts and adaptability helped us endure these tough times—omnivorism helped us weather scarcity.

VITAMIN SEA

Both Marean and Tattersall agree that the Sapiens hanging on in southern Africa couldn't have lived entirely on shellfish. Most likely they also spent time hunting and foraging roots inland, making pilgrimages to the sea during spring tides. Marean believes coastal cuisine may have allowed a paltry human population to hang on until climate change led to more hospitable terrain. He's not entirely sold on the idea that marine life was necessarily a driver of human brain evolution. By the time we incorporated seafood into our diets we were already smart, our brains shaped through millennia of selection for intelligence. "Being a marine forager requires a certain degree of sophisticated smarts," he says. It requires tracking the lunar cycle and planning excursions to the coast at the right times. Shellfish were simply another source of calories.

Unless you ask Michael Crawford.

Crawford is a professor at Imperial College London and a strident believer that our brains are those of sea creatures. Sort of.

In 1972, he copublished a paper concluding that the brain is structurally and functionally dependent on an omega-3 fatty acid called docosahexaenoic acid, or DHA. The human brain is composed of nearly 60 percent fat, so it's not surprising that certain fats are important to brain health. Nearly fifty years after Crawford's study, omega-3 supplements are now a multi-billion dollar business.

Omega-3s, or more formally, omega-3 polyunsaturated fatty acids (PUFAs), are essential fats, meaning they aren't produced by the body and must be obtained through diet. We get them from vegetable oils, nuts, seeds, and animals that eat such things. But take an informal poll, and you'll find most people probably associate omega fatty acids with fish and other seafood.

In the 1970s and 1980s, scientists took notice of the low rates of heart disease in Eskimo communities. Research linked their cardiovascular health to a high-fish diet (though fish cannot produce omega-3s, they source them from algae), and eventually the medical and scientific communities began to rethink fat. Study after study found omega-3 fatty acids to be healthy. They were linked with a lower risk of heart disease and overall mortality. All those decades of parents forcing various fish oils on their grimacing children now had some science behind them (Kromhout, 1985). There is such a thing as a good fat.

Recent studies show some of omega-3s' purported health benefits were exaggerated, but they do appear to benefit the brain, especially DHA and eicosapentaenoic acid, or EPA. Omega fats provide structure to neuronal cell membranes and are crucial in neuron-to-neuron communication. They increase levels of a protein called BDNF, which supports neuronal growth and survival. A growing body of evidence shows omega-3 supplementation may slow down the process of neurodegeneration, the gradual deterioration of the brain that results in Alzheimer's disease and other forms of dementia (Külzow, 2016). Popping a daily omega-3 supplement or, better still, eating a seafood-rich diet, may increase blood flow to the brain (Amen, 2017). In 2019, the International Society for Nutritional Psychiatry Research recommended omega-3s as an adjunct therapy for major depressive disorder (Guu, 2019). PUFAs appear to reduce the risk and severity of mood disorders like depression and to boost attention in children with ADHD as effectively as drug therapies (Chang, 2019; Dyall, 2015; Derbyshire, 2018).

Many researchers claim there would've been plenty of DHA available on land to support early humans, and marine foods were just one of many sources. Not Crawford. He believes brain development and function are not only dependent on DHA but, in fact, DHA sourced from the sea was critical to mammalian brain evolution. "The animal brain evolved 600 million years ago in the ocean and was dependent on DHA, as well as compounds such as iodine, which is also in short supply on land," he says. "To build a brain, you need these building blocks, which were rich at sea and on rocky shores." Crawford cites his early biochemical work showing DHA isn't readily accessible from the muscle tissue of land animals. Using DHA tagged with a radioactive isotope, he and his colleagues in the 1970s found "ready-made" DHA, like that found in shellfish, is incorporated into the developing rat brain with ten-fold greater efficiency than plant- and land-animal-sourced DHA, where it exists as its metabolic precursor alpha-linoleic acid. "I'm afraid the idea that ample DHA was available from the fats of animals on the savanna is just not true," he disputes. According to Crawford, our tiny, wormlike ancestors were able to evolve primitive nervous systems and flit through the silt thanks to the abundance of healthy fat to be had by living in the ocean and consuming algae.

For over forty years, Crawford has argued that rising rates of mental illness are a result of post-World War II dietary changes, especially the move toward land-sourced food and the medical community's subsequent support of low-fat diets. He feels omega-3s from seafood were critical to human's rapid neural march toward higher cognition, and are therefore critical to brain health. "The continued rise in mental illness is an incredibly important threat to mankind and society, and moving away from marine foods is a major contributor," says Crawford.

University of Sherbrooke physiology professor Stephen Cunnane tends to agree that aquatically sourced nutrients were crucial to human evolution. It's the importance of coastal living he's not sure about. He believes hominins would've incorporated fish from lakes and rivers into their diet for millions of years. In his view, it wasn't just omega-3s that contributed to our big brains, but a cluster of nutrients found in fish: iodine, iron, zinc, copper, and selenium among them. "I think DHA was hugely important to our evolution and brain health but I don't think it was a magic bullet all by itself," he says. "Numerous other nutrients found in fish and shellfish were very probably important too and are now known to be good for the brain."

Marean agrees. "Accessing the marine food chain could have had a huge impact on fertility, survival, and overall health, including brain health, in part, due to the high return on omega-3 fatty acids and other nutrients." But, he speculates, before MIS6, hominins would have had access to plenty of brain-healthy terrestrial nutrition, including meat from animals that consumed omega-3-rich plants and grains.

Cunnane agrees with Marean to a degree. He's confident higher intelligence evolved gradually over millions of years as mutations inching the cognitive needle forward conferred survival and reproductive advantages—but he maintains that certain advantages like, say, being able to shuck an oyster, allowed an already intelligent brain to thrive. Foraging marine life in the waters off of Africa likely played an important role in keeping some of our ancestors alive and supported our subsequent propagation throughout the world. By this point, the human brain was already a marvel of consciousness and computing, not too dissimilar to the one we carry around today. It's reasonable to assume it helped us endure a perilous time. "Once we were able to access the coastal food chain in Africa—far more rich and reliable than inland sources of fish—we exploded across the planet," says Cunnane. "After being in Africa for many millions of years we were able to get all the way to Australia in just 80,000!"

Among the migrants who left Africa between 200,000 and 70,000 years ago were those who hugged the coast. They wound along the edge of India and Southeast Asia, eventually reaching Australia with the help of land bridges and crude rafting skills. To this population, the sea would've been an important source of food. "I don't think any one particular ingredient played much of a role in shaping later hominid brain evolution," says Dean Falk. "I think it could have been true for certain local populations in certain habitats. And maybe seafood was important for some coastal groups. But plenty of fish were also available inland."

While certain marine nutrients like omega-3s may have been especially healthy for the hominin brain, many scientists feel that by the time our ancestral brain had evolved to the point where it could march us across the planet, overall calories were far more important to brain function and survival than specific nutrients. We simply needed energy to survive.

This need explains our modern sweet tooth and taste for unhealthy foods. In nature, ripe fruit is one of the only sources of sugar. When you're a starving hunter-gatherer and you come across a plum tree, it's in your best interest to eat as many as you can (and set a few aside for your family). Sugar meant survival. The life expectancy of a Paleolithic human was thirty years at best. Diabetes down the road wasn't a concern.

"Frankly I think it was overall calories that was most important," adds Wrangham. "Other animals have big, intelligent brains and you don't see any signs that gorillas are doing anything special to find omega-3s. Getting them from fish may have been more efficient, but humans got a lot of fat from plants as well. It was mostly about calories."

There is evidence that other hominins consumed seafood. Neanderthal fossils from Eurasia show they suffered from high rates of swimmer's ear, probably the result of marine foraging. And *Homo habilis* remains have been found next to those of catfish. Yet like Falk and Wrangham, I tend to believe hominins got their nutrients and calories wherever they could. If we lived inland we hunted. Maybe we speared the occasional catfish. We sourced nutrients from fruits, leaves, and nuts. A few times a month, those of us near the coast enjoyed a feast of shellfish.

WHAT TO EAT

Whether we're looking at what we ate this morning or what our ancestors consumed 200,000 years ago, interpreting dietary research is difficult. Nutrition studies are often correlational in design. Say people who start each morning with a cup of green tea or a turmeric supplement also have a lower risk of Alzheimer's disease (which research has shown they might). Does this mean green tea and turmeric prevent brain disease? Maybe. But not necessarily. Correlation doesn't prove causality. Maybe green tea drinkers are also more likely to maintain a healthier lifestyle in general. They eat right all around. They don't smoke. They exercise. They've made it to level five on their meditation app.

Studying a single nutrient can come with a lot of confounding variables. This is partly why modern medicine is so prone to flip-flopping. Doctors and nutritionists vilify fats one decade and embrace them the next. Red wine is good for us in moderation. Or maybe it's not. There are so many possible variables in our diets and lifestyles, it's incredibly difficult to parse out what's healthy and what's not if looking at a single food or nutrient.

Vitamins and supplements are now a $40 billion industry. And the vast majority of it is bunk. Or at least not based on credible science. The FDA regulates supplements as food. So unlike drugs, they're not subjected to rigorous efficacy and safety testing. This means manufacturers can more or less claim whatever they want—presuming, of course, their products don't cause people to keel over and die. A 2019 meta-analysis collected data from 277 trials and nearly one million patients and found even widely established vitamins—B6, vitamin A, multivitamins, antioxidants, iron—had no effect on mortality or cardiovascular health, including heart attack and stroke. Yet virtually every bottle of big-brand multivitamins boasts it "Promotes Heart Health" or some such claim. According to the study, even brain-healthy omega-3s have limited evidence supporting their use when it comes to mortality and cardiovascular disease. This isn't to say certain vitamins aren't good for us, especially in subpopulations of people who might be lacking one vitamin or another. It's just, for certain claims and indications the evidence isn't there. Omega-3s seem to have clear benefits for the brain, but it's unclear if one of them is preventing stroke. The marketing is miles ahead of the science.

Another 2019 analysis comparing supplement use with mortality saw similar findings (Khan). Use of dietary supplements was *not* associated with decreased mortality—but vitamin A, vitamin K, magnesium, zinc, and copper *were* associated with reduced mortality and cardiovascular disease when obtained through diet. In a blog post, National Institutes of Health director Francis Collins commented on the results, saying, "These findings serve up a reminder that dietary supplements are no substitute for other evidence-based approaches to health maintenance and eating nutritious food." Advising people to consult the objective health reviews put together by the United States Preventive Services Task Force, he wrote, "Those tend to align with what I hope your parents offered: eat a balanced diet, including plenty of fruits, veggies, and healthy sources of calcium and protein. Don't smoke. Use alcohol in moderation. Avoid recreational drugs. Get plenty of exercise."

It seems much of the familiar, intuitive health advice we've heard for years still holds true. There's something inherent to a healthy, well-balanced diet that, at least for now, science can't pack into a pill. Which is why, amid all the inconclusive evidence and counterclaims, many nutrition scientists have moved to focus on overall dietary patterns rather than individual nutrients. In general, regions of the world with high longevity and lower rates of dementia tend to have similar a diet—one centered around vegetables, whole grains, and seafood, while allowing for occasional meat. Universally, the healthiest human diets (based on mortality rates) are also low in sugar and processed ingredients (Chen, 2019).

The idea of focusing on food patterns for brain health really got going in the early 2000s, when author Dan Buettner published a *National Geographic* cover story on what he and a team of scientists called Blue Zones, regions of the world where people live for an unusually long time. Costa Rica. Okinawa. Sardinia. Greece.

The Okinawan diet is especially intriguing. Despite being an island, its denizens eschew seafood for the most part, instead subsisting mostly on vegetables and leaning heavily on orange and purple sweet potatoes. These local staples are packed with nutrients and fiber and don't raise blood sugar levels to the degree yams and white potatoes do. Okinawans also practice moderation. Before meals, families recite the adage *hara hachi bu*, a reminder to stop eating when they are 80 percent full. Low-calorie diets have been shown to reduce inflammation and the risk of Alzheimer's disease. They also seem to improve mood and mental health. Their most obvious benefit is that they help us lose weight, which comes with its own plethora of mental and physical benefits.

The Seventh Day Adventist community in Loma Linda, California is also a Blue Zone. Its residents adhere to a biblical diet, which, like other diets on the list, is rich in nuts, grains, vegetables, and, for those who aren't strict vegetarians, fish. And, Swedish meatballs aside, the Scandinavian diet is among those thought to promote health, including that of the brain. High in fish, fruits, nuts, and vegetables, the "New Nordic Diet," as it's called, is associated with improved cardiovascular disease risk and reduced stroke risk. Whether it's Atkins, keto, or paleo, fad diets come and go. But certain general dietary philosophies and regional traditions consistently seem to confer brain health.

For many years now, the diet that has best balanced health with a Western palate comes from countries skirting the Mediterranean Sea. The Mediterranean Diet is a generalization for traditional diets you might find in Spain, Italy, Greece, and the Middle East. These cultures tend toward whole grains, healthy fats like omega-3s, and antioxidant-rich fruits and vegetables. Mediterranean diets are high in leafy greens and colorful food—oranges, purples, reds—always a good sign. In nature this is a nutritious spectrum, usually indicating vitamins and antioxidants.

One of the leading proponents of leveraging diet for better brain health is Felice Jacka, a professor at Deakin University and the University of Melbourne in Australia and founder of the International Society for Nutritional Psychiatry Research. The list of dietary data culled from Jacka's work is long. She was one of the first researchers to associate

Western diets high in processed food and excessive poor-quality meat with depression, anxiety, and reduced brain volume. A study published by her and her colleagues in September of 2015 reported that people who consumed a Western diet for over four years had a significantly smaller left hippocampus per MRI scans. Other work by her group assessed the diets of over 20,000 mothers during pregnancy—the children of mothers with the unhealthiest perinatal eating habits had the highest rates of behavioral and emotional problems (Redman, 2018).

Jacka's work parallels other research showing diets high in sugar can result in inflammation and a metabolic domino effect that impairs brain function and contributes to disorders like Alzheimer's disease. Her findings consistently point to more traditional diets as being the healthiest for the brain, in particular the Mediterranean diet, but also those of other fish-fixated regions like Japan and Scandinavia. "Stress and other uncomfortable emotions can cause us to reach for the biscuit tin—they don't call them comfort foods for nothing!" she admits. "But the data consistently show that the main constituents of a healthy-brain diet include fruits, vegetables, legumes, nuts, fish, lean meats, and healthy fats such as olive oil."

Multiple studies report a link between adhering to a Mediterranean-style diet and reduced depression risk. One study found people consuming the MIND diet—a hybrid of the Mediterranean and the high-nutrient, low-salt DASH diet—are 7.5 years cognitively younger than those who don't.

The trendy Paleo diet—which prescribes a diet our Paleolithic ancestors might've eaten, rich in meat, fruits, vegetables, nuts, and seeds—is an anthropological mess. Today's hefty, domesticated cows are a poor nutritional substitute for aurochs, their far leaner ancestors. Not to mention, the Paleolithic age spans 2.5 million years, during which our diet fluctuated as we learned to cook and access new food sources. One thing we definitely didn't do during Paleolithic times was press olives, yet many Paleo advocates allow for olive oil and other plant fats. To their credit though, most interpretations of the Paleo diet are low in sugar and cut out processed food, a pattern that may protect against metabolic syndrome—the dangerous combination of high blood pressure, excess fat, high blood sugar, high cholesterol, and inflammation that can lead to heart and brain dysfunction.

Jacka acknowledges the limitations of correlational research. "It's difficult to tease out cause and effect," she says. But the sheer volume of data that's now been collected on diet and brain health suggests the two are closely related.

Scientists hopefully won't have to rely on correlation for long. A more telling scientific method is the randomized-controlled trial: take two groups of people and randomly assign them to receive a medical intervention. Many RCTs include a control group who, unbeknownst to them, receive a placebo therapy. As of this writing, four randomized trials looking at the relationship between dietary changes and mental health have been completed. Jacka was a coauthor on the first, the SMILES trial. The study found dietary counseling encouraging a healthy, Mediterranean-style diet lowered the risk of depression. In the not-too-near future, diets may become prescriptions. Food as medicine.

During periods of scarcity, overall caloric intake was the prime concern for early humans. The agent of our survival. But the fact that certain dietary patterns appear to be healthier for the modern human brain than others implies they would've been for our ancestors as well. We didn't necessarily need fresh oysters and brightly colored, nutritious fruits to survive—and we still don't. But those who found them might have had an advantage. Those oysters I witnessed being shucked at the American Psychiatric Association meeting aren't enough to ward off dementia or depression on their own. But if incorporated into an all-around healthy diet high in nutrients and low in sugars, they may help.

Our brains evolved at sea, assembled from the same nutrients and building blocks as those of other animals. Yet in primates and, later, hominins, selection gradually favored intelligence, allowing us to adapt to an ever-changing environment and navigate new terrain. We started eating more meat. We harnessed fire. We cooked our food. We exploited the land, the woods, the sea. Our omnivorism provided enough reliable nutrition for evolution to shape our super-sized brain.

CHAPTER THIRTEEN

THE CREATIVES

In Japan, some crows have taken up the practice of using motor vehicles to crack their walnuts.

They grab a nut from a nearby tree and place it in a crosswalk while traffic is stopped. Then they fly off, wait for the light to turn green, and return to a meal once the walnut is crushed, hopping among the pedestrians crossing the street. Researchers think the birds learned how by watching cars crushing nuts that had fallen from roadside trees. The behavior is new, first observed on a college campus around 1990.

If we define creativity as the ability to imagine, or to develop original ideas, humans are not alone in creative spirit. Other species exhibit all sorts of clever behaviors that parallel human innovation. Japanese macaques make snowballs. Dolphins carry around pieces of sea sponge to guard their snouts from sharp coral while fishing. And chimpanzees and bonobos are veritable toolsmiths, with their sticks, rock hammers, and anvils. Jane Goodall described a chimp named Mike who rose to alpha status by banging two kerosene cans together to scare off the competition. Barbara King says chimps are capable of fashioning tools and using them in a sequence to solve complex problems. A wild chimp in the Congo named Dorothy was seen using one club to strike a beehive, another smaller club to open it up, and a slender twig to extract the honey. Ivory Coast chimpanzees hone their hunting skills for ten to twenty years in a kind of creative training process. As primatologist

Christophe Boesch observed in 2002, at around age six they'll begin to approach colobus monkeys, only to run away screaming as adult monkeys scare them off. Within a few years the more aggressive young males conquer their fear and begin chasing down the smaller colobus. By age ten they're usually effective hunters and the smaller colobus don't have much of a chance. Once seasoned, they're not only able to anticipate the movements of their prey but also how the movements of other hunting chimpanzees will influence those of the monkeys being hunted.

It's intuitive—innovative animals are more durable. Research shows bird species that live in changing environments have larger brains compared to their body size than birds in more stable ecosystems. Not having reliable access to resources means they have to get creative when it comes to finding food or material to build nests. Migratory birds are less creative than those that stay local year-round. Unable to endure a chilly winter, they have no choice but to fly south. Small-brained birds, less able to learn the dangers of their environment and adapt, are more likely to get hit by cars (Laland, 2017; Sayol, 2016; Moller, 2017). The Japanese crows, on the other hand, have large corvid brains and are very aware of traffic patterns.

The cleverness of other apes in particular tells us the early hominin brain was wired for creativity long before we scored symphonies. It was preadapted to an inventive revolution. Some anthropologists argue that selection for creative intelligence was among the most important drivers of human brain evolution and is what enabled innovations like taming fire in the first place. Selective pressures acted on creative pursuits in ways that made us better suited to our environment. Becoming upright freed up our hands to help actualize ideas requiring dexterity. Invention helped us survive and intrigue potential mates.

At some point in our evolution, while ancestral chimps were using sticks to collect insects—which is, by most species' standards, an incredibly intelligent behavior—we crossed a threshold from highly instinctual animals to highly cognizant, symbolic creatives. We became crafters, builders, artists. Creativity led to practical contributions like axes, butchering tools, and weapons. Our big brains could free-associate; new ideas came quickly, as did symbolic language and thought. We began engraving, building bug-resistant bedding, and sewing. We started banging drums, painting cave walls, painting ourselves.

In his 2017 book *The Creative Spark*, University of Notre Dame anthropologist Agustín Fuentes writes, "Creativity is at the very root of how we evolved and why we are the way we are. It's our ability to move back and forth between the realms of 'what is' and 'what could be' that has enabled us to reach beyond being a successful species to become an exceptional one."

Unfortunately, the earliest evidence of human ingenuity is probably lost to history and decomposition. Poking sticks and spears rotted away. But stone tools and fractured animal bones found in both Kenya and Ethiopia suggest Australopithecines like Lucy were using tools to break the bones of prey and butcher meat over three million years ago, well before *Homo* arose. The Stone Age had begun, and continued until just 7,000 years ago when we started working with metal, marking the dawn of the Bronze Age. The first tools known to have been used by humans were unearthed from Tanzania's Olduvai Gorge in the 1930s by British anthropologist Louis Leakey, whose work at the site helped prove humans evolved in Africa. Leakey's findings show that beginning around 2.6 million years ago, both *Homo habilis*, or "handy man," and *Homo erectus* were chipping stones into crude hand-axes, crafting our first cleavers and knives. This was the Oldowan tool industry and the first human foray into mass-production. Thousands of Oldowan tools and tool shards have since been found across Africa, the Middle East, Europe, and Asia (Harmand, 2015).

Two factors essential to human Paleolithic survival were our abilities to hunt and cook. As genetic mutations inched our brains toward higher complexity—and with our increasing taste for meat—making tools became increasingly important for protection, hunting, and breaking down carcasses. Early hominins more adept at such skills no doubt enjoyed a safer, more nutritious lifestyle. It's not a coincidence that during our two-million-year tenure as toolmakers, the hominin brain tripled in size. But a question among anthropologists—much like the chicken-or-egg question around meat consumption—is whether tool-making actually helped drive human brain evolution, or was instead one of many byproducts of a large, intelligent brain.

At Emory University's Paleolithic Technology Laboratory, anthropologist Dietrich Stout and his students not only attempt to re-create ancient tools, they scan each other's brains while doing so. Stout believes toolmaking was the initial spark that urged our brains down a highly creative evolutionary path. He wrote in a 2016 *Scientific American* article that natural selection would have favored "any [genetic] variations that enhance the ease, efficiency, or reliability of learning the new trick." The better stonemasons would've brought added protection to themselves and their community; this would've helped them survive and find a partner. Genes conferring more innovative toolmaking and fine motor control became our genomic norm, since people without tool and weapon-crafting abilities would've been outcompeted pretty quickly, their genes lost for good.

Stout's lab is a trade school for making both Oldowan axes and the more sophisticated Acheulean tools that came along a million years later. Named for Saint-Acheul in France where they were first discovered, Acheulean tools are thinner and more finely honed into sharp blades. Stout teaches his students to craft these early tools as our Paleolithic ancestors would've, by knapping, or using a hammer rock to chip sharp shards off a stone core. Working with neuroscientists, he then analyzes what areas of the brain are most active while they knap. Employing a brain-imaging technique called FDG-PET—short for fluorodeoxyglucose positron-emission tomography—his team found that both Oldowan and Acheulean knapping activate a brain region called the supramarginal gyrus. This is a fold in our parietal lobe involved in the spatial perception of our body and limbs. Chipping away at Acheulean-style tools also activates a part of the prefrontal cortex involved in response control and decision-making. Stout uses another imaging technology called diffusion tensor imaging, or DTI, to map the brain's white matter (all those myelin-coated axons that allow neurons to talk to each other). His DTI research shows knapping increases the

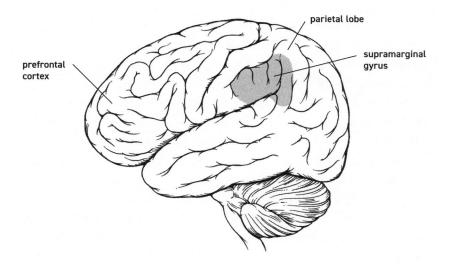

connectivity between the same frontal and parietal regions that light up under the PET scanner. The degree of connectivity correlates with how many hours each knapper has spent training. Stout believes advanced toolmaking requires more mental control and the ability to plan ahead, and that having a brain capable of these abilities would've been highly selected for. Ape research supports his claim. DTI scans show toolmaking circuitry is far more developed in humans than in chimpanzees. Chimps can bang two cans together to exert dominance, and as Jane Goodall observed, they can strip away leaves from twigs to make more-effective insect-gathering tools—but they can't hone a hand ax to a razor-sharp edge. That requires a degree of planning and patience only hominins have ever enjoyed.

In his 1950 book *Man the Tool-Maker*, English anthropologist Kenneth Page Oakley wrote that toolmaking was the "chief biological characteristic" that drove human evolution. Other apes, he concluded, "are capable of perceiving the solution of a visible problem, and occasionally of improvising a tool to meet a given situation." But to "conceive the idea of shaping a stone or stick for use in an imagined future eventually" is beyond their mental capacity. Stout explains that this once-popular idea fell out of favor when we observed other animals like chimpanzees, crows, and octopuses using tools. When Jane Goodall first reported tool use among chimps, Louis Leakey responded, "Now we must redefine tool, redefine Man, or accept chimpanzees as human."

The social brain theory is also partly responsible for demoting "man the toolmaker" from evolutionary prominence. More recent ideas like Robin Dunbar's make a solid case that social species tend to have larger brains and that our social behavior and intelligence was responsible for the rapid enlargement of the hominin brain. In a way, research like Stout's unites both ideas. Much as Dunbar believes the social intelligence hypothesis to be entwined with how we interacted with our environment, early toolmaking would've been a social pursuit. Stout and others believe mimicry and the ability to teach and learn toolmaking skills were integral to toolmaking's importance. It's even been proposed that because toolmaking relates to the human ability to communicate through gestures, it may have been an evolutionary precursor to language, in that it was a vehicle for mimicry. The exchange of information about our creations among individuals is what made innovation such a powerful adaptive force. We were dads at Home Depot comparing cordless drills.

"From the days of our earliest, protohuman ancestors, we have survived and increasingly thrived because of our exceptional capacity for creative collaboration," writes Fuentes, who thinks many of the integral moments and advances in the human story—social complexity, harnessing fire, making stone tools—involved our hominin ancestors creatively collaborating. Creativity turned ideas into the tangible. As we became more social and developed language, our ideas and talents could be easily shared and passed on to the next generation. As Fuentes puts it, "Two million years ago our small, naked, fangless, hornless, and clawless ancestors with a few sticks and stones surmounted near impossible odds. All because they had one another and a spark of creativity. And so do we."

University of St. Andrews evolutionary biologist Kevin Laland concurs, writing of early human creativity, "It also gave them the intelligence to invent new solutions to life's challenges and copy the innovations of others. This favored further brain expansion in an accelerating cycle that climaxed with the evolution of human cognition."

All of us primates have disproportionately large cortices and, as a result, are prone to innovation. Laland believes the only reason monkeys and chimpanzees don't have their own tool industries is their inability to share information. Our capability to communicate and compare notes sets us apart from all other species and allows ideas to spread through communities. "The absence of complex culture in other animals isn't down to a lack of creativity," Laland says. "Rather it's their inability to transmit cultural knowledge with sufficient accuracy. That's why no monkey ever composed a sonata."

PORTRAIT OF AN ARTIST

Tools and weapons offer obvious benefits when it comes to hunting, protection, and sourcing food. Both would have been hugely advantageous to our ancestors, particularly as we became more verbal and social and capable of sharing and passing down knowledge. The pace of our creative progress, however, was painfully sluggish. If we assume the Oldowan-type hand ax dates back at least three million years, we have to wait another million just to sharpen the thing, and still another million before a hominin thought to tie a pointy rock to a stick to create a spear.

It wasn't until we became truly symbolic that the course of humanity rapidly changed. Tattersall believes the ability to creatively communicate through language and other forms of symbolism started us on a course of relentless creativity, which hasn't slowed since. "Radical innovations happened very infrequently. But once you've got symbolic *Homo sapiens* on the scene, we go from one generation of technology to the next at an increasingly fast pace," he says. "In the old days, hominins would adapt existing tools to new purposes as environmental niches and demands changed. But we started inventing new tools for new purposes. This is a completely different way of dealing with information in the brain. Information is now exchangeable, and can be passed on to the next generation."

American linguist Daniel Everett makes the case that as early as 350,000 years ago *Homo erectus* appreciated tools with some sense of symbolism. Even a simple hand ax would've had many uses and elicited cultural memories of shared hunts and meals. "*Erectus* tools," he writes, "represented culturally agreed-upon meanings that referred to *displaced*—not immediately present—activities and meanings, the hallmark of symbols." Perhaps crafting and appreciating tools seeded some appreciation of an object apart from just the outcome it enabled. We first cracked open gazelle bones to enjoy the marrow, then we came to appreciate the ax that cracked them.

Early symbolic humans began assigning symbols to represent things, including intentions, people and places, or other animals. An extended index finger came to mean "look over there," not "look at my finger." A certain sound could mean "zebra," and another, "mammoth." A painting of a gazelle could represent the horned beast without actually being one.

Another major signal of human cognitive advance was playing with a powdery clay called ochre, which refers to any iron-rich rock that can be molded and used as yellow, orange, and red pigments. Anthropologists think we originally used ochre as a sunscreen or insect shield. But a human face or stone wall smeared with the stuff became too intriguing to waste on practical purposes. There would've been plenty of crumbly rocks of other colors available in the Middle Pleistocene, yet fossils show we were most intrigued by reddish hues, as our languages still reveal today.

While many languages group colors like blue and green together, most have a dedicated word for red. Some only have words for *red* or *not red*. Using ochre as paint or makeup may have been one of our earliest forms of symbolic communication, which makes evolutionary sense. Recall that, around twenty-three million years ago, primates evolved a third type of retinal cone, which allowed us to see red. Trichromatic vision helped us better navigate arboreal life and locate ripe, brightly colored fruits. Monkeys, apes, and humans are exceptionally aware of red, especially against the tans and greens of a savanna or woodland. The color of blood and mortality may have weighed heavy on the symbolic brain.

It's impossible to say what our ancestors were initially conveying by processing ochre for symbolic purposes. It was our original medium for decorating rocks and caves. On our faces, it may have helped establish tribe identity or scare other groups off. Pieces of ochre intentionally shaped by humans date back over 300,000 years. But it's those found at Pinnacle Point, from around 160,000 years ago, that, according to Curtis Marean, convincingly suggest symbolic behavior. As part of his research into our marine eating habits, Marean and his team have uncovered newer, more extensively processed ochre samples. The clay has clearly been worked and ground into a paste or paint with an intended use. Marean believes symbolic cognition may have already been in place in southern Africa as modern humans branched from the hominin tree some 200,000 years ago. Pinnacle Pointers were among the first humans to consistently forage the shore and may also have been among the first artists.

For around sixty-five million years, we primates existed at the mercy of our genomes. Many, if not most, of the evolutionary forces that shaped our brains, along with ecological influences and just plain happenstance, acted through our genes, as mutations here and there conferred added cognitive abilities and creativities adaptive for survival. Whether it was advances in hunting, creative thinking, or simply peeling fruit, nearly all our qualities and creativities trace, in part, back to mutations giving certain ancestral individuals survival and reproductive advantages. And then, around 90,000 years ago, our genes got left behind to a degree. We were preadapted for bigger things, with our cognitive wiring exploding into what we would call culture. Without much genetic change to our brains at all, and armed with symbolic thought, language, and social skills, the pace and influence of cultural change and innovation raced past that of our genome. We started sharing more knowledge than ever before, trading with other communities, and banding into larger and larger groups. Humans became a species of Edisons and Teslas, constantly tinkering. Yuval Noah Harari—a history professor at the Hebrew University of Jerusalem and author of the massively successful book *Sapiens*—calls it our "cognitive revolution."

Among the oldest known works of art are two pieces of ochre engraved with a cross-hatched design. Found at Blombos Cave in South Africa, about 60 miles west of Pinnacle Point, they date back at least 70,000 years and demonstrate their maker had clear symbolic intent. Maybe scoring lines in clay amused the artist or their campmates. Maybe with plenty of cooked food on the fire, they were just bored, killing time doodling. The ochre art doesn't appear to have served any practical purpose other than artistic expression for the sake of it.

By 40,000 BC, and probably much earlier, we'd become a species of artisans, inventing arrowheads, awls, and fishhooks. Beads made from shells became our jewelry and cave walls our canvases. Upper Paleolithic paintings exist at hundreds of locations across Europe and Asia and will surely continue to be discovered. Some of the oldest known visual works, discovered at three different sites in Spain, were painted by Neanderthals—proving that Sapiens weren't the only human artistes.

The first documented *story* might be a 44,000-year-old painting found in an Indonesian cave. It depicts a buffalo being hunted by human-like forms with tails and beaks wielding spears. The oldest sculpture and three-dimensional depiction of a human is the "Venus of Hohle Fels," carved from mammoth ivory and found in 2008 in a cave near Schelklingen, Germany. Venus figurines are voluptuous depictions of women that were all the rage throughout late Paleolithic Europe; they're thought to have been either part of symbolic rituals or, more likely given their often exaggerated proportions, the first pornography (Hoffman, 2018; Henshilwood, 2018).

As language and vocalizations took on tones and melodic qualities that originally helped communicate information but later became enjoyable, the human voice was surely the original musical instrument.

Squirrel monkeys can distinguish sound patterns much in the way humans can melody, suggesting the primate brain was primed for responding to tonal patterns. Anthropologists think some appreciation of percussion also came early in ape evolution. A 2019 study by researchers from Kyoto University in Japan added to evidence that chimpanzees respond to rhythm. They clap their hands, swaying and tapping their feet to piano music. Early humans probably had a similar ear for infectious beats, eventually realizing auroch, elk, and buffalo hides could make excellent drumheads if you stretched them tight enough. Two bone flutes dating from 35,000 years ago were also found at the Hohle Fels cave. These, along with flutes made the wing bones of mute swans discovered in France and Austria, are among the earliest known musical instruments (Hattori, 2019).

Over 300 caves containing early human art have now been found in France and Spain alone, including the famed sites at Lascaux and Altamira, where ochre and other pigments were used to paint humans, horses, aurochs, and boars. Artwork at Chauvet Cave in southern France depicts lions, bears, hyenas, and a disembodied vulva hovering beneath a bison head. Things were getting weird.

THE ART OF BELIEVING

It's difficult to say whether or not our creative pursuits in the past 100,000 years were biologically selected for. Probably to some degree. But socially transmitted behaviors and information came to dominate our lives. As Richard Wrangham writes, "Culture is the trump card that enables humans to adapt, and compared to the two-million-year human career, most cultural innovation has indeed been recent."

Steven Pinker argues that the mind evolved specific systems he calls "modules" to solve the challenges of survival. Think of them as circuits that connected brain regions for abilities like understanding math, language, and identities. Knowledge and intelligence would've been an adaptation to better navigate our world, but, at its extremes, creativity wasn't necessarily selected for. We didn't evolve to write novels and paint landscapes. More likely, literature and the arts are nonadaptive abilities that arose as byproducts of practical skills like language and communication. Visual art came from a brain already highly perceptive to color and responsive to the reds and purples of plums and pears. Pinker famously called music "auditory cheesecake, an exquisite confection crafted to tickle the sensitive spots of . . . our mental faculties." We enjoy sweets not because we evolved a taste for confection, but because we developed a palette for the sugars of ripe fruit and the rich fats of nuts, seafoods, and meats. The arts are pleasures, akin to drugs and dessert. "Cheesecake packs a sensual wallop unlike anything in the natural world because it is a brew of megadoses of agreeable stimuli which we concocted for the express purpose of pressing our pleasure buttons," Pinker writes, likening dessert to the dopamine-driven, reward center rush of porn and other visceral hobbies.

Others argue visual art and music are old enough to have had an evolutionary influence, probably by encouraging social bonding and group identity. Dunbar believes that, along with laughing and language, song was essential in releasing endorphins, which cemented our social relationships. Pointing to the tonal and lilting languages, Dean Falk believes musicality may have evolved along with spoken word. "There's music in language. In affect and tone of voice, there are songs," she says.

Maybe certain creative pursuits became sugary indulgences that grew from other adaptations more directly related to survival—but creativity has come to define our species. We are eight billion apes capable of poetry and prose and quantum mechanics. Through a union of biological and cultural influences, our creative collaboration built human culture and civilization, in ways both constructive and not. We come together to support causes, while at the same time, starting wars. As Fuentes says, "This collaborative creativity drove the development of religious beliefs and ethical systems and our production of masterful artwork. Of course, it also tragically fueled and facilitated our ability to compete in more deadly ways."

By 100,000 years ago, both Sapiens and Neanderthals had some concept of ceremony and ritual, the roots of spiritualism and religion. Hominin skeletons from about that time found at the Qafzeh and Es Skhul Caves in Israel were buried with artifacts like beads made from shells, deer antlers, and the jaw of a wild boar. By 13,000 BC, caves were being used as what you could call cemeteries. Humans were consistently being put to rest in select locations, often with shell and ivory jewelry. It's hard to say for sure what any of this means and what was going through the cortex of the early human who reddened a set of antlers with ochre and mounted it on a stick at the entrance to a gravesite. It seems symbolic in some way, possibly a tribute to the dead. Or maybe simply a signpost.

The oldest known church or place of worship in human history is a site in Turkey called Göbekli Tepe, which dates back at least 10,000 years. Ruins and clay sculptures found here and at nearby sites suggest that by this point, we'd devised more formalized rituals and religions. And plenty of phallic representations. And more curvy Venus figurines.

Anthropologists consider the earliest spiritualism to have been an amalgam of different views on animism, from the Latin *anima*, meaning spirit or life. As Harari writes, it's the general idea that animals, geographies, and objects have a spiritual meaning and essence. Early ideas of spiritualism must have taken countless forms as we developed meaningful social connections and fantastical ideas inspired by our physical surroundings and perplexing cosmos. Animism was a kind of prelude to theism, the belief in a god, or more often, multiple gods. We came to believe in all-powerful beings distinct from mortal humans. By the time monuments and tombs like Newgrange, Stonehenge, and the Egyptian Pyramids were erected 5,000 years ago, we had gods of darkness, dusk, the sun, the moon, the stars, vengeance, knowledge—we concocted gods for just about everything! By this point, we'd also begun burying our dead in increasingly elaborate fashions, and not long after, during the Uruk period (around 4000 to 3100 BC), Sumerians started etching little marks into clay tablets. This was cuneiform, the earliest known writing system, which they used in part to log their religious beliefs.

Harari writes that it was our embrace of collective rituals and ideas that supported the growth of human societies. Civilizations. Chiefs and chiefdoms. Ideologies. Ethical codes. These are all man-made creations that exist because enough of us agree they do. In *Sapiens*, Harari writes that the basis of modern civilization traces back to our embrace of shared myths. We may only be capable of managing 150 meaningful relationships, but civilization and common beliefs allowed us to band into much larger groups, with people we'd never met before and probably never would. Increased cooperation and creativity allowed us to develop abstract entities like societies, spiritual systems, and armies. "You could never convince a monkey to give you a banana by promising him limitless bananas after death in monkey heaven," writes Harari, but fictions like the biblical creation story gave us the "unprecedented ability to cooperate flexibly in large numbers." Harari uses modern businesses as another example of a shared myth. The French car company Peugeot, he writes, employs over 200,000 people and produces 1.5 million cars a year. Yet it only exists insomuch that enough people and a legal system, itself a manmade entity, agree it does. "If a judge were to mandate the dissolution of the company, its factories would remain standing and its workers, accountants, managers, and shareholders would continue to live—but Peugeot would immediately vanish," he writes. "Peugeot seems to have no essential connection to the physical world . . . [it] is a figment of our collective imagination."

So why do we believe in entities that don't exist? Why is doing so inherent to so many of our cultural and religious traditions?

We are social, creative animals and most of us want to belong. We gravitate to like-minded cliques. We seek popularity among our peers. Belonging to a group was adaptive in that it afforded support and protection. If an especially charismatic alpha male told you that last night a thunderous yellow sun god with eyes of fire spoke to him, and you must worship that sun god or face the consequences, you'd better go along with the story, lest you risk your acceptance among the group (or worse, your life at the hands of the alpha). The tale may seem a little suspect, but show some enthusiasm and you're good. You're protected. Going along with the fantastical is innate to human culture and has served us well.

Hyenas and wolves will band together to scavenge and hunt, and chimps will gang up to raid their neighbors, but no other species forms long-term mass-cooperatives bound together by shared beliefs like humans do. By believing in our grand myths, we were able to cooperate with thousands and eventually millions of strangers in our towns, states, and kingdoms. Our creativity granted us the social and cultural fabrications that still hold society together today.

CHAPTER FOURTEEN

THE NEW WAVE

For most of human existence, we lived in roaming, ragtag bands, hunting, gathering, and later making temporary camps. Some of us remained in Africa while others spread throughout Europe and Asia. Toward the end of the glacial age, roughly 18,000 years ago, an intrepid group of humans journeyed across a land bridge that connected Russia to Alaska at the time and is now submerged beneath the Bering Strait. These were the Paleo-Indians, the ancestors of modern Native Americans. Within just a few thousand years, they'd made it all the way to southern Chile. *Homo sapiens* had become a truly global species.

It could've been climate change, or population growth, but 10,000 years ago, something got us thinking we should plant seeds. Between the Old and New Worlds, agriculture arose independently at least eight different times, rewriting our way of life, many believe for the worse. Corralling livestock and cultivating crops may have provided more reliable sustenance, as well as reserves during lean times. In China, we grew soybeans and rice. In the Levant, part of the Fertile Crescent, it was barley, flax, lentils, and chickpeas. Pigs were domesticated in multiple parts of Asia and later, in Europe, crossed with wild boars. Both the Olmecs and the Mayans in Mexico domesticated and grew maize. We'd become agriculturalists. And shortly thereafter, what we think of as modern civilizations begin to appear, with culture being shared faster than ever through denser and denser populations.

Michael Tomasello proposes the idea of cultural learning as a uniquely human form of social learning. As civilizations grew, humans began conforming to and identifying with cultural groups, which in turn expected people to conform to them. Children were taught by their parents they were part of this culture or that. Tomasello wrote, "These special forms of cultural learning enable powerful and species-unique processes of cumulative cultural evolution." By the agricultural revolution, we had the mental ability to learn skills and information and share them with high fidelity within the culture with which we identified. And what a disaster it was.

Relying on domestication and farming was a risky move. The plains and woodlands were an omnivorous cornucopia, whereas agricultural societies rely on just a few species of cultivated plants and animals. Droughts and blights can do a lot more damage on a farm than in the biodiverse wild—and they did. Genetic data show at least two massive dips in the human population as we moved our farms into central and northwestern Europe (Shennen, 2013). Yet overall, farming proved more productive than hunting and gathering, and the human population skyrocketed as a result.

Sedentary farming brought with it many ills of modern society. More people meant more crowding and spread of disease. Agriculture enhanced the concept of ownership. Being able to control and distribute a resource obsessed us with land, production, trade, and capital, the basis of modern economies. This led to positions of power and influence and increased class division—and with them dictators, tyrants, and taxes.

American anthropologist Jared Diamond believes agriculture was a catastrophe for gender roles, cementing the expectation that women stay home and keep house. Women in farming societies typically had more children and performed more demanding work than those living as hunter-gatherers. Scholars think before we had farming, many human groups were matrilineal, meaning kinship was traced through the female line. As territory and ownership became more important to us, human society shifted power and inheritance to the paternal side (with plenty of notable exceptions such as Judaism). The physically dominant sex wanted control of their crops and land, and sadly, of women. Exodus 20:17 famously warns against coveting "thy neighbor's wife." In *Civilized to Death*, Christopher Ryan suggests considering the entire passage: "Thou shalt not covet thy neighbor's wife, nor his manservant, nor his maidservant, nor his ox, nor his ass, nor any thing that is thy neighbor's." Wives had become property.

Agriculture began the long human tradition of social stratification, which, at its worst, enslaved people into servitude. A 2019 study published in *Science* suggests that as far back as 4,000 years ago, Bronze Age humans of different social classes were living together. The researchers analyzed DNA from more than 100 ancient skeletons to determine relatedness among people living on ancient farmsteads in what is now Germany. Related individuals were laid to rest with goods and belongings that appeared to be passed down through generations. The unrelated people in the household were buried with nothing, suggesting they were considered a lower class and not given ceremonial treatment.

By looking at the radioactivity profiles in teeth from the skeletons, the study authors were also able to determine where each person grew up. They found nearly all households studied included females who hailed from elsewhere. The remains suggest farmsteads were passed through many generations of males, whereas females only persisted in their communities for one generation. A system of patrilocality was in place, in which men stayed in their place of upbringing, while women moved in with their husband's family. Patrilocal cultures had previously existed, including far back in the Paleolithic, but these findings support the idea that the practice became more common as organized farming societies developed.

The social setup could've been the natural outcome of men using their egos and physicality to stay put. But there are other arguments for why patrilocality came to be. A woman marrying outside her community would've increased a population's diversity, preventing the genetic problems that come from inbreeding. The practice may also have encouraged the exchange of cultural information and trade; increasing social interactions with other communities allowed for more-efficient transfer of skills, goods, and farming tips to a wider population.

In their 2010 book *Sex at Dawn*, Christopher Ryan and Cacilda Jethá argue that prior to agriculture and the concept of private property, humans were dedicated polygamists. Much like bonobos, we freely shared sexual partners. It may not have been the gleeful orgy from Bosch's, *The Garden of Earthly Delights*, but more promiscuity did help keep the peace through ambiguous paternity. Bonded through frequent sexual encounters, everyone chipped in to raise the children collectively. In agricultural society, patrilineal jealousies strengthened a commitment to direct inheritance. Women were now bound to men, who were intent on their land and property being passed on to their biological heirs—"No way I'm leaving my flax farm to *that guy's* kid." Some argue our migration toward monogamy was also a result of our large brains and extended childhoods—children with two dedicated parents might've fared better—and that it reduced conflict within a group by helping avoid sexual jealousy.

The rapid changes to our social lives, sex lives, politics, and economy brought on by agriculture were largely the result of cultural change and shifting behaviors and attitudes, not biological evolution. Harari cites priests, popes, and Buddhist monks as evidence of the degree to which culture overcame biology. Alpha male chimpanzee males try to mate with as many females as possible to propagate their selfish genes; many modern religious figures, including priests, the "Catholic alpha males," as he calls them, commit to celibacy out of tradition. Biologically it doesn't make much sense, and yet the Catholic church has survived for 2,000 years.

Agriculture may have been a disaster for biodiversity, women's rights, and egalitarianism, but over just a few thousand years, it became the norm. The human condition migrated toward reliable calories, power, and politics. "There was no way for us to foresee the consequences of our initial success. We simply took what was given us and continued to multiply and consume in blind obedience to instincts inherited from our humbler, more brutally constrained Paleolithic ancestors," writes E. O. Wilson.

GENUS, GENOMES, GENIUS

Leaving cultural evolution aside for a moment, many of the evolutionary forces that shaped our brains acted through our genes. The right mutation here and there increased our brain size, the number of neurons it contained, and how they connected in ways that make us different from other species. Our DNA probably plays some role in just about everything we say, think, feel, and do. Whether its advances in hunting, creative thinking, or simply peeling fruit, nearly all of our qualities trace back in part to mutations that, along with environmental influences and luck, gave certain individuals survival and reproduction advantages over others. This can be a dicey acknowledgment, one that strays uncomfortably close to eugenic idealism, especially when applied to qualities like intelligence and creativity. How much of *us* is due to nature (our genes) versus nurture (our environment and culture). Are we born with talent, or do we acquire it through practice? The argument over where our skills come from goes way back.

Victorian polymath Sir Francis Galton—who coined the phrase "nature versus nurture"—founded the nineteenth-century eugenics movement. He noticed certain skills and qualities run in families, and he believed the population could be "improved" through selective coupling. As history attests, this sort of thinking is near-universally toxic to human society. Other thinkers over the years—including psychologist K. Anders Ericsson and the writer Malcolm Gladwell—have taken an opposing view. While acknowledging that genes play a role in certain skills, they believe rote practice—the "10,000-hour rule" as Gladwell calls it—can allow us to achieve success at just about anything.

Research by psychologists David Z. Hambrick and Elliot Tucker-Drob looking at musical abilities in twins found it's hard to separate nature and nurture. Success at a particular pursuit may rely on how our genomes interact with our environment. The genetic influence on musical achievement is far greater in those who practice more. Repetition seems to exponentially build on a biological predisposition for musicianship; the influence of genes becomes more important the more we practice. Famed jazz pianist Thelonious Monk taught himself to read music as a child without a single lesson; by age thirteen he was barred from taking part in the Apollo Theater's amateur competition because he'd won too many times. This was probably the result of intense practice bringing out his natural abilities.

"The nature vs. nurture debate is—or certainly should be—over," says Hambrick. "It is no longer productive to think of expertise as 'born' versus 'made.' We must embrace the idea that both are important—and that their influences are entwined. Practice and training can activate genetic factors." Hambrick adds that while there must be a minimum amount of training to become skilled at something, we have no idea what that minimum amount would be. "Babies don't emerge from the womb knowing how to execute a perfect jump shot . . . and people may differ massively in how much training it takes them to reach a certain level of skill."

I could almost end this book right here. By the time Sapiens were dining on cultivated lentils and forcing their neighbors to harvest them, the modern human brain had arrived. Rare geniuses aside, if you control for environmental factors like poverty and education, most humans are similar when it comes to intelligence. The range of smarts among us is nearly negligible compared to the range between humans and the next most intelligent species, chimps and bonobos.

But the geneticists impel me to keep going. Some genomic researchers claim to have identified not only genes that helped distinguish our intelligence from that of other apes but also those that led to meaningful changes in our brain not long after the agricultural revolution. These findings are not without controversy.

Two genes in particular, microcephalin and the abnormal spindle-like microcephaly associated gene (*ASPM*), are thought to directly contribute to brain size—children with mutations in either gene are born with a cerebral cortex similar in size to that of early hominins. Research by geneticist Bruce Lahn found both genes underwent rapid evolution in humans following our split with chimpanzees (Gilbert, 2005). This implies the random occurrence of certain variations in their DNA sequence offered our ancestors a significant survival advantage. But it's worth noting that when Lahn and his colleagues reported that one *ASPM* variant showed up over 6,000 years ago—just as agriculture and civilization really began flourishing—and was more common in humans living in what are now Europe, the Middle East, North Africa, and parts Asia, he was met with accusations of practicing racist science, while critics poked holes in his research methods, partly discrediting his claim. Cultural evolution was paramount in our recent history. But whether or not Lahn's findings hold water, natural selection may still have played a significant role at certain times in our modern brain development, albeit in a way heavily entwined with our culture and ecology.

In 2017, a team of American and European researchers identified an additional fifty-two genes associated with human intelligence, which they felt represented only a small fraction of genes involved in our abilities to reason and problem solve. A year later they added 939 more genes to the list. Overall genetic influence on intelligence grows from around 20 percent in infancy to over 60 percent in adulthood, meaning, much like musical ability, our experiences draw on and amplify genetic inclinations (Sniekers).

Genes, of course, don't just lend our brain competence—they hinder it as well. Countless gene variants have been linked to just about any condition you can think of, including brain disorders like depression, Alzheimer's, bipolar disorder, and autism. In 2016, a gene called *C4* was found to be a major risk factor for schizophrenia (Seker). Normally, *C4* controls a process called synaptic pruning, which eliminates redundant or unused neural connections in the brain during our teen years. After birth and through mid-to-late childhood, the human brain typically amasses trillions upon trillions of synapses—far more than there are stars in the Milky Way! Half these connections are long gone by adulthood. Immune cells called microglia literally nibble them out of existence if they're not needed. When *C4* is mutated, the synapses instead stick around, and the excess connections result in scattered brain activity, including the hallucinations, delusions, and confused thoughts that come with psychosis. All told, well over 100 genes have now been linked with schizophrenia.

Synaptic Pruning at Work

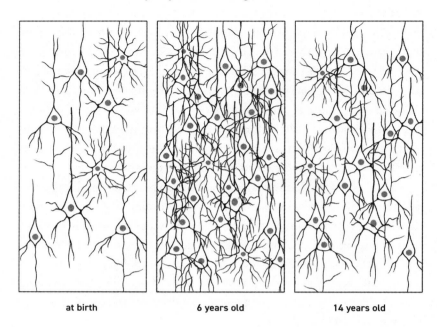

at birth   6 years old   14 years old

You'd think natural selection would've weeded out mental illnesses, but there are arguments for why they've been conserved through evolution. Some psychiatric conditions might be pathologic exaggerations of beneficial traits. A little anxiety has clear evolutionary benefits in avoiding danger, but a little too much is crippling. The manic periods that come with bipolar disorder can inspire extreme ambition and creativity, as they may have done for Vincent Van Gogh, Virginia Woolf, Kurt Cobain, and a long list of other artists thought to have suffered from the disorder. Psychiatrist Nassir Ghaemi argues effective leadership is often a result of mental illness, with depression instilling empathy and realism, and mania reinforcing resilience, and creativity. He proposes it was depression that allowed Lincoln, Martin Luther King, and Gandhi see their empathetic visions through; and a manic episode that led Sherman to march through Georgia.

The human brain evolved to be so creative that, today, it creatively malfunctions. Take schizophrenia. As far as we know, we are the only animal to suffer from psychosis, the risk for which is largely influenced by genes. In 2015, geneticist Joel Dudley published research showing certain gene variants that contribute to schizophrenia are involved in higher-order thinking, driven by our prefrontal cortex (Xu). An incredibly complex network of genetic influences and neuronal interactions contribute to human cognitive abilities. Perhaps when the wiring is tangled or our neurotransmitters unbalanced, there's simply more that can go wrong—complex function begets complex malfunction.

With our big brain came the perils of having one.

CHAPTER FIFTEEN

FUTURE SAPIENS

Aristotle believed the heart to be the seat of our soul and the brain just a radiator to dissipate heat.

Writing 500 years later, Greek physician, surgeon, and philosopher Galen partially corrected things when he proposed that our sensations, thoughts, and memories were manifestations of the brain, with the heart and liver handling our personalities and emotions. He believed we have three souls, the brain being one of them. He also believed the brain is formed from sperm. Medical science can move slowly.

Two thousand years later, we're able to slap electrodes on our scalps and read our brain waves. Neurosurgeons drill holes through our skulls and snip out brain tumors and little pieces of misfiring tissue that cause seizures (while we're awake!). Yet compared with every other organ, the human brain remains a scientific mystery. A neurology professor once told me our understanding of the brain is where our grasp of the heart was in the 1950s. And I bet even this estimate is generous.

A major question among scientists is where the human brain is headed. Is it still evolving to any great degree? As genetic engineering technologies improve, will we artificially evolve our brains? Will our new digital world change our neurologic function in enduring, meaningful ways?

The short answer to two of those questions is yes.

As our culture blitzkriegs by, natural selection on the human species is still occurring—only at a much slower rate than cultural change. A period of rapid selection began about 40,000 years ago and lasted through the agricultural revolution as we spread across the planet, encountering new climates, diseases, and cognitive challenges. As some populations migrated north to areas with limited sunlight, there were adaptations to skin color and cold temperatures. As we lived in more crowded communities surrounded by livestock, there was selection related to pathogen resistance and immune function. With the introduction of dairy into our adult diet, there was selection to produce lactase to digest lactose. As recently as 10,000 years ago, there were evolutionary changes in genes controlling neurotransmitter function. Even as our culture went whizzing past our genomes, our gene frequencies continued to change as we adapted to more congested agricultural life (Wang, 2006; Hawks, 2007).

"Selection is still happening in all sorts of ways," says Wrangham.

The most obvious shifts in our collective genome are a result of mass casualties. As I finish writing this book, the world is battling the COVID-19 pandemic. The combination of genetic influences that leave certain people and populations more vulnerable to the virus with environmental factors like socioeconomics, will certainly influence the collective global genome. Events such as mass infections, genocides, and holocausts erase a significant chunk of humanity's DNA and shift the human genetic profile.

These modern examples have the same effect as Neolithic males of one clan killing off those of another. The authors of a 2015 study were perplexed when their data suggested that between 5,000 and 7,000 years ago, the effective population size of men across Africa, Europe, and Asia fell to the point where they were outnumbered by women seventeen to one. By sequencing the Y chromosome of nearly 300 males from 110 different populations they found that, around that time, male genetic diversity had plummeted. Could there have been climate shifts, infections, or other ecological factors that only affected males? Maybe, but seventeen-fold is quite a drop in numbers.

Troubled by how this could be, Stanford geneticist Marcus Feldman reasoned it wasn't that most men died off, but that they fought each other's genomes out of existence to the point where entire male lineages were wiped out. Only the victors' Y chromosomes survived. Male genetic diversity, not their overall headcount, sank, as a smaller majority of male lineages wielded all the power and survived.

I'd like to believe de Waal's poetic view that civilizations were formed, in part, through our maternal and peaceful side—call it our inner bonobo. But I'm not sure the genetics agree. If Feldman's interpretation of the data is correct, the violent alpha males and promiscuous dictators held the most leverage over humanity's future. Eight percent of the men who now live in what was once the Mongol empire, after all, share a Y chromosome tracing back to Genghis Khan.

Wrangham returns to the idea of self-domestication. In recent human history, execution and incarceration have been the leading means of controlling violence within a society, which changes reproductive rates and skews gene frequencies away from those contributing to aggression. Modern medicine also drives some degree of evolution by saving lives that would've been lost to disease in Paleolithic and Neolithic times.

Smarts are always under selection pressure, says Christof Koch, president of the Allen Institute for Brain Science in Seattle. "Just look at the Darwin awards!"

The tongue in cheek honor has been around since 1985 and awards those who "eliminate themselves in an extraordinarily idiotic manner, thereby improving our species' chances of long-term survival." The rhino poacher killed by an elephant. The Utah man who disposed of dynamite by shooting it with a shotgun. The young man on a Boston booze cruise who perished after doing a handstand on the railing of the ferry. Is making light of their deaths problematic? Probably (though maybe not for the rhino poacher), but there is data showing a clear relationship between intelligence and mortality. In 1947, over 65,000 children in Scotland took part in an intelligence survey measuring abilities like following directions, reasoning, mathematics, and understanding proverbs. In 2017, researchers published a study looking at who among the lot was still living in 2015, and the cause of death in those who'd died. They found those with higher IQs were far less likely to have died of heart disease, stroke, certain cancers, and lung disease. There was a linear relationship between smarts and longevity, with exceptionally intelligent people living longer than the very smart, and the very smart living longer than people with average intelligence (Calvin).

The obvious critique is that those from more affluent families and with more education make more informed lifestyle choices, like not smoking, and have more access to healthcare. And research regularly correlates socioeconomic status with better performance on intelligence tests. But the Scottish study and others controlled for socioeconomics and still found a strong correlation between intelligence and mortality.

Twin studies show IQ is heritable to some degree, but at the moment it's impossible to say by how much. So much of our cognition is shaped by experience, family, support, and encouragement. Intelligence is a product of genes and environment. And especially of the environment acting on genetic predilections.

Migrations and cultural forces that influence who we have children with will forever alter our genetic profile. As humans move around the world for whatever reason—a new job, an industry boom in one region of the country, escaping persecution—genomes evolve with new breeding patterns. Typically these changes are too small to notice in just a few generations. Although perhaps not always.

According to University of Cambridge psychologist Simon Baron-Cohen, in the case of techie geographies like Silicon Valley, we may see rather rapid genomic change. Brother of the actor Sacha Baron-Cohen (a.k.a. Borat), he shares a similar hubris, proposing that people drawn to technical fields like computer science, machinery, and mathematics have a higher degree of autistic traits. As tech centers become localized, like-minded partners pair up and have children. As a result, he posits, local genomes take on more autistic tendencies, which are then passed on through generations. These are regional micro-evolutions.

Whereas past research has tended to focus on the social impairments that characterize autism, there is a movement to embrace the incredible attention to detail and technical abilities that often come with it—to depathologize the condition. Modern technology is hugely important to today's society. If Baron-Cohen is right, Northern California could see future generations of skilled computer engineers like never before, as a result of evolutionary change (Baron-Cohen, 2006).

Still, biological evolution tends to be such a slow process, it's difficult to see in real time, especially in large populations. "Cultural evolution is so much more rapid now," says Koch. "If you were to take a modern baby and somehow time travel it back 2,000 years, or vice versa, I don't think either of our cultures would notice much difference between the children." Two thousand years equates with less than 100 human generations, a blip in our evolutionary story. Pinker says selection is always active, but the human population is far too large, diverse, and moving in too many cultural directions to know where the human brain itself is moving.

INFLUENCERS AND INHERITANCE

Some scientists think a big part of where our brains are headed may depend on "epigenetics," the concept that changes to our DNA acquired through life can be passed on to our children. In a way, this idea reopens a centuries-old scientific debate.

Over 200 years ago, early evolutionary thinker and French naturalist Jean-Baptiste Lamarck proposed that changes to life-forms could be acquired throughout life and passed to the next generation. If a giraffe extended its neck to reach the leaves of an acacia tree its neck would grow, and its offspring would inherit a longer neck. Darwin came along and countered with the now-accepted theory of evolution—giraffes evolved longer necks by adapting to their environment over generations. He had no idea what DNA was. No one did at the time. But modern genetics helped solidify Darwinian dogma as the most likely explanation for how we all got here. Giraffes didn't acquire longer and longer necks due to their behavior, but to their genomes. Those that had acquired random mutations coding for a longer neck were more likely to survive, outcompeting their shorter peers by reaching the loftier leaves.

Epigenetics in no way resurrects Lamarck's theory of the ambitious giraffe, but the idea that biological change not coded for by DNA can be inherited has an ethos of Lamarckism about it. Epigenetic changes include modifications to proteins called histones, around which DNA wraps, and to DNA methylation, essentially the binding of a molecular cap that prevents certain genes from being expressed. There is a long list of lifestyle influences that may exert epigenetic changes, everything from diet to toxins to childhood anxiety. Drinking, obesity, and poor sleep are also factors. Exposure to these elements through life may modify sperm and egg cell DNA in ways that stick around.

A well-known series of Swedish studies looked at how access to food among a generation born in the early 1900s in a small community called Överkalix influenced the future health of their children and grandchildren. The results, and those of subsequent studies, found higher rates of mortality due to heart disease, diabetes, and cancer in men whose grandfathers had experienced a good farming season between ages nine and twelve. Something related to being well fed during development may have changed the eventual grandfathers' biology in ways that were passed along.

Fair warning: the self-help sensationalism around epigenetics precedes the science by a mile. Meditation and stress reduction are good for us in myriad ways, but scientists don't yet know if this has to do with an epigenetically modified double helix, as some people claim. However, many scientists do believe that, in the long-term, epigenetic effects could tinker with our brains in permanent ways. Cunnane believes the modern world will see a confluence of epigenetic influences that alter the brain. Our increasingly poor diet, high in processed foods and low in nutrients and seafood, "might really have a mass epigenetic effect on our cognitive abilities." He also worries our new hyperconnected yet inward culture will gradually change our brains as we stare at our smartphones and communicate virtually. "We don't socialize or talk to each other as much. We don't look at each other. This will certainly have an effect. Maybe through epigenetic modification."

With hesitation, Koch says social media and modern technology may be affecting young brains in some endurable way, including making them more anxious. Dean Falk goes much further, firmly believing technology is actually speeding up biological brain evolution. The same was true, she says, when we began reading 5,000 years ago, and more recently during the personal computer revolution of the 1980s.

Social interaction was among the most important drivers in our brain evolution. Now, virtual connections like social media give us the impression we're socializing, but at the expense of actual human interaction.

Enough Instagram likes could mean we're not alone. We belong.

Maybe that's fine. Instagram and Snapchat might provide the same essential flow of information and gossip we've indulged in for millennia. However, sociologists worry the new communication is cultivating a culture of anxiety and loneliness, while at the same time convincing us we're accepted. We're increasingly self-obsessed, yet somehow less self-reflective and more cognitively isolated. We're distracting ourselves from our real selves. Whether this will have epigenetic effects remains to be seen, but assuming our new digital behavior is passed down culturally for generations to come—and we can only imagine how much more virtual our lives will get—will this have any serious detriment to our species?

Christopher Ryan writes that, based on psychology research, "the single most reliable predictor of happiness is feeling embedded in a community." Census data show that in 1920, 5 percent of Americans lived alone. Today over a quarter do, the highest percentage ever recorded. Human societies have always been built on community, just as they are among our monkey and ape relatives. Now technology has imprisoned us in earbuds.

Perhaps this is why rates of mental illness are soaring. We are a social species depressed in isolation.

Even dating apps could be an evolutionary force. As we know from trolls and vitriol we see online, people behave differently behind the digital wall. Today, someone too shy to make a connection at a bar, who happens to be a riot on Tinder, now has a chance at a date; or the other way around, as charm doesn't always translate to 500-characters, curated bios, or texting. Partnerships that never had a chance can now form through a new type of interaction, and others that might've been won't get a chance.

"Yeah, maybe."

Tattersall isn't entirely buying my theory.

"So much of all of this is speculation. When you look around at modern kids, you say there has to be some consequence. What the nature of that consequence will be I have absolutely no idea." Tattersall thinks of recent human history as a sloshing bucket of water, with gene frequencies going this way and that but tending to stabilize. "I'm a great believer in reversion to the mean, and I don't think things are going to change as much as we think. It's easy to look around and say, 'wow, this time the shit is really hitting the fan.' But it's so hard to predict."

"So," I asked, "are we getting dumber?"

"We're not getting dumber! I worry much more about climate change."

STAYING HUMAN

Neuroscience titan and 1906 Nobel Prize–winner Santiago Ramón y Cajal said, "Every man can, if he so desires, become the sculptor his own brain."

Today, that's a concern. Science could advance to the point where we use it to steer our own brain evolution.

Genetic engineering techniques now allow scientists to literally edit people's genomes—to remove unwanted genes and splice in favorable ones. This could usher in a welcome new era of therapy for genetic disorders, but also a host of inevitable eugenic pursuits. It's easy to imagine designer babies—and with them, designer brains—aren't too far off, their embryonic genomes edited for desired traits.

The gene-editing technology that's generated the most buzz these past few years is called CRISPR-Cas9. It's a mouthful—short for "clustered regularly interspaced short palindromic repeats and CRISPR-associated protein 9." Adapted from a process that naturally occurs in bacteria as a defense against invading viruses, CRISPR allows scientists to recode our genomes by adding, deleting, or altering DNA at specific locations. They make a piece of RNA in the lab that attaches to both a targeted region of DNA and the Cas9 enzyme, which cuts the DNA. A customized DNA sequence can then be spliced into the snipped genome using the body's own DNA repair machinery.

To date, most gene-editing research has focused on preventing and treating diseases known to be caused by certain genetic mutations, like hemophilia, cystic fibrosis, and certain forms of epilepsy. It's also being looked at in more genetically complex conditions, including cancer, heart disease, and mental illness, all of which involve a host of genes interacting with environmental risk factors.

CRISPR and other genetic technologies mostly focus on somatic cells, or cells that don't pass on genetic material to the next generation. Alter a sperm or egg cell (our germline cells) or an embryo, and you've changed a reproductive lineage forever. This is where things could get dicey. Theoretically, technologies like CRISPR could be used to enhance skills, strength, and intelligence. It doesn't feel like much of a stretch to imagine this ability, in the hands of a mad scientist or eugenic dictator, leading to clandestine cyber caves where super embryos are grown to service powerful armies and infrastructures.

Germline editing is illegal in many countries. Though, as of this writing, it's being pursued in Russia and has been used to create genetically modified babies by at least one, now infamous, researcher. His name is He Jiankui, a Chinese scientist who, in 2018, announced he'd used CRISPR to create the first gene-edited babies, twin girls named Nana and Lulu. He claimed to have modified a gene called *CCR5* in order to protect them from contracting HIV from their HIV-positive father. Never before had a human embryo been genetically edited, implanted into a mother, and brought to term.

He Jiankui first presented his experiment at a gene-editing summit in Hong Kong. Summit chair and Nobel-winning biologist David

Baltimore later said the work should be "considered irresponsible," and that it was a "failure of self-regulation by the scientific community." NIH Director Francis Collins called the announcement "profoundly disturbing." Even CRISPR pioneers were horrified. They felt the science wasn't ready. There was no moral consensus. He had sought ethical advice from other scientists and ignored it. Words like "monstrous" were used.

In December of 2019, He was found guilty of conducting "illegal medical practices" and sentenced to three years in prison. I'd be shocked if, in twenty years, experiments like his aren't commonplace.

Enabled by better stem cell technology, CRISPR is now entering the realm of the brain, bringing us one step closer to mental manipulation through rescripting our DNA. Stem cells are those that haven't yet developed into a particular cell type. In 2006, Kyoto University stem cell researcher Shinya Yamanaka discovered a way to reset adult skin or blood cells into stem cells. Using different cocktails of chemicals called transcription factors that regulate gene expression, he showed stem cells can be coaxed into forming any cell type in the human body. Muscle cells. Liver cells. Immune cells. Brain cells. All of them. The work earned Yamanaka a Nobel Prize.

Scientists can now grow an unlimited supply of any cell type using these induced pluripotent stem cells, or iPSCs, while avoiding the ethical issues of harvesting stem cells from human embryos, which had been controversial. Prior to Yamanaka's revelatory work, it was difficult for researchers to study brain tissue, because they were limited to neurosurgical samples that could only survive a few days; iPSCs made neurons far more available.

At first CRISPR didn't work well with stem cell–generated brain cells. Stem cells repair damaged DNA quickly, making it difficult to employ a technique that relies on breaking it. In 2019, a team at the University of California, San Francisco, developed a CRISPR variant in which Cas-9 is deactivated. This means they can't splice any genes into a strand of DNA, but they can use CRISPR to locate genes involved in brain function and structure and either increase or decrease their expression.

The innovation could be invaluable to the future of understanding and treating brain disorders—it's also one step closer to applying gene editing to our thoughts and intellect (Tian).

If CRISPR concerns weren't enough, 500 miles south of UCSF, at the University of California, San Diego, iPSCs are being used to grow something approaching a brain from scratch.

Scientists have been trying to grow human organs for over a decade. Skin. Guts. Little kidneys and livers. They sit in laboratories, capable of approximating biological functions, but without a body to help maintain. These "organoids" are not fully formed functional organs. They're miniaturized proxies that help researchers model various diseases and test therapies. In 2019, scientists from UCSD revealed they'd successfully grown "mini-brains" with the neural activity of a preterm infant (Trujillo). There'd been earlier efforts to coax stem cells into brainlike collections of neurons, but none had demonstrated brain activity mimicking the real thing. This time, after just two months, the scientists detected scattered brain-wave activity of roughly a single frequency, much like that seen in the immature human brain. By ten months out—when each organoid was about the size of a pea—the mini brains' activity zapped at different frequencies in a predictable pattern, similar to that of the maturing human brain.

Like with CRISPR, there is vast potential here, especially in modeling conditions like schizophrenia, autism, and epilepsy. These organoids also make ideal testing ground for new drugs, giving lab rats a well-deserved break. And their most obvious role may be that of replacement parts, patching bits of brain claimed by trauma or stroke.

If cultivating a copy of our most sophisticated organ—the seat of our thoughts, personalities, and behaviors—in a small plastic dish sounds like a pulp fiction nightmare, I'm with you!

Koch says don't panic.

"To be clear, no one would confuse these with an actual brain. They almost certainly don't feel anything," he says. Brain organoids lack both the neurons for sensing pain and the circuitry for processing it. They don't have a vascular supply so can't grow much larger than a bean.

Petri dish of cultured one-year-old brain organoids, each the size of a lentil

He did, however, say *"almost* certainly."

Size and sensory limitations could change as stem cell engineers learn to cultivate different neuronal subtypes and oxygenate organoids with blood vessels. Could a tree of capillaries bring a laboratory brain to life? What if it did start to feel something? Would it be distressed? Or worse, in agony? And how would we know?

THE PROBLEM WITH ZOMBIES

Artificial intelligence is advancing to the point where we ask machines to order us sweatpants and Keurig coffee pods. It's worth asking whether computers will ever be able to feel, or experience consciousness.

As German philosopher Thomas Metzinger put it on neuroscientist Sam Harris's podcast: "For a number of years, I've argued against even risking phenomenal states in machines. We should in no way attempt to create conscious machines or even get close because we might cause a cascade of suffering."

Consciousness is a tricky concept to define. We all feel we have it, but what exactly is it and where does it come from?

Grasping what it is to be a conscious being has stumped philosophers and scientists for ages. Many thinkers agree it centers around everything we subjectively experience. The song that just popped into our head. The love we feel for our partner. The slice of Neapolitan pizza we're eating. As Harris puts it, "The fact that it's like something to be you, is the fact of consciousness."

The natures of our life experiences are called qualia—these are qualities you can't convey to anyone else. The redness of red. The smell of that pizza. How you experience a song. Qualia are the kind of concept that blow college-student minds ("Dude, what if the blue you see is different from the blue I see? Think about it." *takes bong hit*). Then there's the actual physical world, which, even if we perceive it differently from one another, most of us agree exists.

This gets at dualism, the idea of separate physical and mental worlds. Dualism is most associated with philosopher René Descartes, whose "I think, therefore I am" is perhaps the most familiar one-liner in western philosophy. The only catch is that in Descartes's opinion, "we"—our thoughts, our personalities, our minds—are mostly divorced from our bodies. The polymathic Frenchman and other dualist philosophers proposed that, while the mind exerts control over our physical interaction with the world, there is a clear delineation between body and mind; our material forms are simply temporary housing for our immaterial souls. However, centuries of science argue against a corporeal crash pad. The body and mind appear inextricably linked.

So how can a physical collection of funny-looking cells bound together into three pounds of brown Jell-O create something as marvelous as subjective experience and qualia? In the 1990s, Australian philosopher David Chalmers called this "the hard problem," as opposed to easier neuropsychological queries, like understanding perception, attention, and memory. A related thought experiment invoked among modern philosophers of the mind is "the zombie problem." If an entity can move, speak, and act just as we can, is it conscious? Or could it be relying on a convincing physical system void of feeling and consciousness?

In the movie *Her*, Joaquin Phoenix's character falls in love with an operating system voiced by Scarlett Johansson. Stanley Kubrick's *2001: A Space Odyssey* stars a sentient computer named Hal with ideas of its own. The zombie problem might argue these computer systems only seem conscious.

Then there is the philosophy of materialism, popular among scientists. Generally speaking, materialists believe consciousness, qualia, and feeling can arise from physical systems, like a human brain. They argue specific architectures and circuits in the human brain somehow came together to make us conscious beings.

"This is the standard thinking in Silicon Valley," says Koch, "that if you eventually give Alexa enough upgrades, she'll be conscious." It's a modern version of the Turing Test, computer science pioneer Alan Turing's question of whether or not machines can think. Many materialists believe we'll eventually have a high-fidelity human brain running on a computer. You boot it up, it says hello, it feels and behaves just like us.

Another view is that everything in the world, from mountains, to trees, to mice, has some version of consciousness. This view is called panpsychism; it's the basis for integrated information theory, or IIT. Though he's involved in some of the most cutting-edge neuroscientific research, including brain-computer interfaces, and is perhaps not a likely candidate to believe his potted plant has some degree of awareness, Koch is open to IIT. "How do we know consciousness doesn't extend beyond humans?" he asks. If we define consciousness as the ability to feel something, how do we know single cells aren't conscious? Maybe when the right cocktail of molecules comes together to form a bacterium—the right integrated information—the microbe snaps into some new state of being. Maybe, as John Muir speculated, that microbe is even capable of some modicum of a quality akin to enjoyment. Even single-celled beings and the cells that make up our bodies are so vastly complex that modern science can't come close to assembling one. "Maybe when intact, a cell feels like something, and when it's destroyed, it no longer feels anything," says Koch.

I tend to agree that consciousness must have arisen gradually; and that evolution didn't suddenly just turn it *on* in humans. Elements of consciousness may have appeared early on in animal evolution, perhaps even predating what we think of as animal today. I can't imagine it didn't follow a gradual Darwinian process, in which mutations here and there enhanced the animal experience, changing how we viewed and existed in the world. Subjective experience gradually arose from objective matter. Monkeys and chimpanzees may experience the world more deeply than, say, mice. And *Australopithicus* and *Homo erectus* absorbed life even more.

Descartes felt that a machine capable of uttering words that corresponded with its "bodily actions" was conceivable. But, as he put it, "It is not conceivable that such a machine should produce different arrangements of words so as to give an appropriately meaningful answer to whatever is said in its presence, as the dullest of men can do." Well, Siri and Alexa prove he got that part wrong. But Descartes' second distinction holds more weight: "Even though some machines might do some things as well as we do them, or perhaps even better, they would inevitably fail in others, which would reveal that they are acting not from understanding, but only from the disposition of their organs." Today those organs are circuit boards and computer processors.

IIT asks whether or not consciousness can arise out of a physical system, and whether a system can be greater than the sum of its parts. The idea is close to the philosophical concept of emergence, in which an entity has qualities unique from those its components have on their own—entities like cities, water, and coordinated flocks of airborne birds. Most scientists believe that consciousness is, in fact, an emergent property generated by our neural wiring.

Yet not all are on board. Sam Harris writes, "To say that consciousness emerged at some point in the evolution of life doesn't give us an inkling of how it could emerge from unconscious processes." He floats the idea that consciousness might be a spandrel, what evolutionary biologists call characteristics that are evolutionary byproducts, meaning they're not selected for through natural selection. One example would be the spots and floppy ears that come with domestication, neither of which serve a biological purpose. In speaking with Richard Dawkins on his podcast, Harris says "It's conceivable to me that [consciousness is] not doing anything . . . everything that you're aware of—your thoughts, intentions, and emotions—is all being generated unconsciously."

Dawkins believes consciousness must arise from information processing in the brain. But he concedes, "To quote Monty Python, on this kind of conversation, my brain hurts. I can't think my way through that. I doubt anybody else can. But that's no comfort."

The philosophers, psychologists, and neuroscientists continue to wrestle. One booming voice in the melee is Tufts University philosopher, Daniel Dennett, whose ideas imply consciousness is an illusion conjured up by our brain—that it doesn't exist at all! As for Koch, he doesn't envision man ever fully coming from machine. We won't, he believes, ever be able to design consciousness or upload our digital afterlife to a server in Arizona. "These ideas are super cool, but they are just clever simulations. It's all pretend. It's all fake. Alexa will not one day have real consciousness."

What's interesting as we watch the development of artificial intelligence unfold, is how difficult it's been to mimic some of biology's most basic functions. Computers capable of complex math have been around for decades, yet a robot dog that can barrel through a doggie door and catch itself when it falls is still a ways away. These are innate functions of the brain and nervous system that are tough codes to crack in digital form.

Koch admits it's very hard to know where the brain is going. Many thinkers on such matters aren't overly concerned with technology interfering with our psyches. Scientists will most certainly be tinkering with genes involved in brain disorders and function. But qualities like personality and intelligence or complicated conditions like depression and anxiety involve hundreds if not thousands of genetic influences interacting with our lifestyles and environmental exposures.

Tattersall doesn't think our brains are going anywhere drastic in the foreseeable future. Our population is too large and has too much genetic inertia for much change to occur, he says. "Cultural change is where it's at. This is where we will see the greatest change."

Pinker tends to agree. "Predictions from the 1990s that yuppie parents would soon implant genes for intelligence or musical talent in their unborn children seemed plausible in a decade filled with [genetic discovery]. But today we know heritable skills are the products of thousands of genes, each with a minuscule effect . . . I doubt we'll direct the evolution of our brains any time soon, if ever." Given that modern parents are "squeamish about genetically modified applesauce," he doesn't feel many would take a chance on genetically modified children.

IN THE FUTURE

It's undeniable that when humans show up, we ruin everything.

As members of *Homo sapiens* made their way out of Africa and around the world, we wreaked havoc on other species, especially the megafauna. In Europe and Asia, we extincted over half such species and in Australia over 70 percent. We did remarkable damage in the Americas, killing off over 80 percent of large animal species. Goodbye mammoths and mastodons. So long, giant ground sloths, American camels, 300-pound beavers, and those 2,000-pound armadillos the size of a Volkswagen. The farther we went in our global migration, the more damage we did. Back in Africa, only about 16 percent of large animals were driven to extinction by humans. African fauna coevolved with humans for millions of years. We learned and adapted to each other and knew each other's tricks. But the rest of the world couldn't keep up with the rapid spread of human migration, hunting, cognition, and culture.

We had little awareness of what we were doing. We lived the best we knew how based on our biology, instinct, and cultural creations. If there was available meat, we ate it. We conquered the world through a lethal combination of wits and a relentless use of resources.

Our brains are so adept at exploiting the planet that we've left it a mess, with rising tides and rising temperatures, and without thousands of once-thriving species. In 1962, Johns Hopkins geneticist H. Bentley Glass told the *New York Times* that, were a global nuclear war to break out, only bacteria and cockroaches would survive, battling it out and taking over the "habitations of the foolish humans." Our future will rely on whether we can harness our instincts to avoid ceding our planet to the microbes and insects.

The evolution of the human brain isn't the story of a single influence or trait. It's an interwoven patchwork of selection and culture taking place over millions of years—our diets, our creative pursuits, our friends, our tools, our fire. E. O. Wilson says the human condition is the result of an idiosyncratic prehistory entwined in complex emotions, thoughts, and behaviors that "stamped our DNA at every major step along the way."

We are now bombarded with more information than ever before. It's hard to know what this means for the future of the brain, but we do know we're processing this new sensory deluge with the same old neurological machinery we've had for many thousands of years. One could assume the drastic increase in information flow, along with our increasingly cyber behavior, will have some sort of lasting influence on our neurobiology.

In his 1986 song, "The Boy in the Bubble," Paul Simon sang of "staccato signals of constant information," a somewhat sarcastic celebration of the new digital world. Those Reagan-era signals now seem quaint. They've since jammed together into the single continuous draw of a screeching bow—a never-ending din of sensory input pummeling our now-ancient senses. The sharing of information was crucial to our evolution, arguably the most crucial factor in making us human. But is there a threshold above which our brains become overloaded—or simply distracted—from all the real-life social information and interaction that got us here?

French author Michel Houellebecq's 1998 novel *The Elementary Particles*, features a forlorn biologist with a drinking problem named Michel Djerzinski. Djerzinski's cloning research ends up rendering human sexual reproduction unnecessary and eventually results in the human species replacing itself with a newer, smarter, more compassionate people, void of cruelty, anger, and ego. "The creation of the first being, the first member of the new intelligent species made by man 'in his own image' took place on 27 March 2029," Houellebecq wrote. "Today, some fifty years later, there remain some humans of the old species, particularly in areas long dominated by religious doctrine. Their reproductive levels fall year by year, however, and at present their extinction seems inevitable."

Houellebecq's example is, of course, extreme utopian satire. Hopefully our big brain thinks up new ways to exist on our planet responsibly, while at the same time addressing the damage we've done. Neuroscience will advance. Genomes will evolve. Culture will change by the second. And we may endure, just as our ancestors did 150,000 years ago, huddled in a rocky coastal cave, struggling to crack open an especially difficult oyster before the tide sets in.

ACKNOWLEDGMENTS

I first have to thank my father, Dan Stetka, who passed away as I was writing this book, and to whom I owe, entirely and gratefully, my interest in science and medicine. When I was eight years old he handed me a book on photosynthesis that I carried everywhere for a while, pretending to understand what something called nicotinamide adenine dinucleotide phosphate was. My dad was a career geneticist and invaluable in offering advice and thoughts on human evolution as I prepared my book proposal (less valuable when digging out dusty genetics textbooks from the 1970s).

To my mother, Mary Lou Stetka, who, when hearing I'd signed a book deal, immediately baked my favorite childhood cinnamon rolls and sent them FedEx overnight to my Brooklyn apartment—thank you, Mom!

I also have to thank the long list of biologists and anthropologists and psychologists and physiologists who were gracious enough to share their knowledge and time with a first-time book author: Detlev Arendt, Jordi Paps Montserrat, Jakob Vinther, Neil Shubin, Alexandra DeCasien, Michael Tomasello, Frans de Waal, Richard Wrangham, Brian Hare, Martin Surbeck, Steven Pinker, Katerina Semendeferi, Barbara King (William and Mary!), Dean Falk, Curtis Marean, Michael Crawford, Stephen Cunnane, Drew Ramsey, Felice Jacka, Joel Dudley, and Christof Koch.

Thanks to Ian Tattersall for a fascinating behind-the-scenes tour of the American Museum of Natural History in New York City; and to the Center For Great Apes sanctuary in Wauchula, Florida, for all they do to provide homes for chimpanzees and orangutans rescued or retired from circuses and research labs; as well as to my mother-in-law and father-in-law, John and Linda Petrusich, for all their support and enthusiasm, and for putting up with a three-hour tour of the Center For Great Apes sanctuary in Wauchula, Florida.

I need to thank my college pal and esteemed ecologist John Grady—a.k.a. Mr. Mesotherm—who donated countless hours bantering with me about evolutionary and ecological theory over the past few years.

I suppose I should thank Charles Darwin for making much of this book possible.

Thanks to my agent Erik Hane at Headwater Literary Management for approaching me about this project in the first place, and to my editors Will McKay and Jacoba Lawson at Timber Press for making it a reality. All three of you have been a pleasure to work with, and incredibly helpful.

To my longtime colleagues at Medscape.com and WebMD.com, where I've been kindly employed for fourteen years. And to NPR and *Scientific American* for their support over the years (a few passages in this book were adapted from my past work at *SciAm*).

I'd like to thank my cat, Carl, for repeatedly walking across my keyboard as I was finishing this book, in an obvious act of literary sabotage.

Finally, I thank my wife, Amanda Petrusich, who supported me through this whole process and who has taught me more than anyone about writing, narrative, and how to handle the occasional waves of terror that come from putting a piece writing out into the world. She is my partner in everything. And she was mostly okay with a book called *Chimpanzee Politics* sitting on our coffee table for nearly a year.

NOTES

CHAPTER ONE: VERY APE

[Page 21] Jane Goodall was the first anthropologist to report witnessing a chimpanzee "fish" for insects. In October of 1960, she observed a chimp named David Greybeard poking pieces of grass into a termite hole and then raising them to his mouth. She realized he was waiting for the termites to bite into the grass so he could eat them. He was making termite popsicles.

[Page 22] Hahn and colleagues determined gene duplications and deletions by examining genomes and analyzing gains and losses in gene number over the millennia. Among humans and chimps, they found 1,418 duplicates that one or the other species doesn't have, suggesting that changes in the copy number of certain genes was a major driver in mammalian and primate evolution (Demuth, 2006). It appears there was a burst of gene duplications in the primate lineage around the time of our split with chimpanzees; both chimps and humans have far more duplicated genes than our orangutan and macaque cousins (Marques-Bonet, 2009). While gene duplications are more often neutral, and in many cases harmful, they are thought to be an important mechanism for driving genomic novelty and evolutionary change (Magadum, 2013).

CHAPTER TWO: LIFE FROM NO LIFE

[Page 29] Though Darwin gets much of the credit, British naturalist and Darwin contemporary Alfred Wallace independently proposed the theory of evolution by natural selection. This spurred Darwin to collect his ideas and writings, which he'd kept secret for almost twenty years, and publish his ideas jointly with Wallace in 1858.

[Page 31] The process of one organism living inside of another, usually out of mutual benefit, is called endosymbiosis. It was initially theorized in the early 1900s by Russian botanist Konstantin Mereschkowski and substantiated in the 1960s by evolutionary biologist Lynn Margulis.

[Page 35] Among our many neurotransmitters are serotonin and dopamine, on which we can blame our depressions and gambling addictions (the former influences our mood, the latter our feeling of reward). Over 200 neurotransmitters have now been identified, with varied functions in the brain and nervous system. These are chemical messengers that transmit information from one neuron to another, or from a neuron to a muscle cell. Some, like glutamate, acetylcholine, and norepinephrine are excitatory, activating neighboring neurons. Others, including serotonin and GABA, decrease synaptic transmission. They act as neuronal brakes. The idea that disorders of the brain are often due to altered neurotransmitter levels—or a "chemical imbalance"—is probably an oversimplification, but the molecules do play an important role in brain function.

[Page 35] Ion channels are also involved in heart function, muscle contractions, insulin release, and maintaining cell volume, literally helping prevent them from exploding or shriveling up.

[Page 35] It's possible that all bilateral animals evolved a central nervous system from a common ancestor, which then branched off into the myriad brain varieties we see today. It's also possible, given the evolutionary advantages of having one, that the nervous system arose independently many times. This remains a major question in evolutionary biology.

[Page 36] Neurons could have evolved from any number of cell types. Most bilateral animals develop from embryos made up of three cellular layers, the endoderm, mesoderm, and ectoderm—or the inner, middle, and outer layers. Some argue neurons arose from epithelial cells, an ectodermal cell layer that lines our skin, gut, and blood vessels, and the layer from which most bilaterian neurons derive. In 2019, Detlev Arendt proposed two other candidates as the neuronal precursor cell. Neurons, he believes, may have evolved from either mesodermal cells—some of which look like modern neurons—or choanocyte-like cells.

CHAPTER THREE: FISH, HEAD

[Page 41] Some scientists believe sea lilies also represent a missing link between vertebrates and invertebrates. They are echinoderms, the phylum that includes starfish, sea urchins, and sea cucumbers. Sea squirts, in the phylum Chordata, are a candidate too; though they're sessile as adults, their larvae look like tadpoles and have a supportive notochord.

[Page 41] Prior to oxygenation increasing Earth's quantity of limestone, many rocks were rich in a mineral called dolomite, which crystallizes slowly. More oxygen encouraged the formation of limestone, which contains aragonite and calcite—calcite being a calcium source from which bone can mineralize. Both minerals could've induced bone formation much faster and with less energy than dolomite.

[Page 46] Goffinet's 2017 article in *Development* nicely reviews the changes in brain structure that began after synapsids split from a common ancestor with diapsids, or birds and reptiles. Unlike birds and reptiles, early synapsids were evolving essential features of the mammalian brain, including a more layered cortex and a tendency to fold and involute to create more surface area.

CHAPTER FOUR: LIZARD KINGS AND LEMURS

[Page 64] At a 1965 "Evolving Genes and Proteins" symposium Zuckerkandl and Pauling outlined their theory that proteins can help determine evolutionary divergence among species.

CHAPTER FIVE: UPRIGHT CITIZENS

[Page 79] Linnaeus was an eighteenth-century Swedish botanist and physician who formalized our system of classifying organisms. He is known as the father of modern taxonomy, the study of classifying life.

[Page 85] The announcement that Neanderthals were responsible for the cave art at three different Spanish sites came prior to the fossil evidence showing that *Homo sapiens* made it to Greece 200,000 years ago. At the time of this writing there is no evidence Sapiens were in Western Europe 60,000 years ago, when the drawings date from.

[Page 89] Even as this book was going to print, advances in our understanding of primate brain evolution came to light. A study published in Science in June 2020 looked at a gene called *ARHGAP11B*, known to be involved in expanding our cortex during development. When the gene was inserted into marmoset embryos, researchers found that after around three months, the monkeys had larger neocortices with more folds, suggesting it's a critical genetic factor in what distinguishes the human brain from that of at least some other primates. By the time you read this, many more such discoveries may have been reported.

CHAPTER SIX: GROOMSMEN

[Page 101] Some birds—especially ravens and crows, or the corvids, as well as parrots—have disproportionately large brains (or brain regions rather) and exhibit an impressive degree of sociality. Some live in complex social networks. Others maintain monogamous relationships through life, its own form of social complexity, requiring a lifetime of processing social queues and behaviors from a single partner, and reacting accordingly. Both lifestyles support the social intelligence hypothesis. The smartest of birds may not be writing novels, but their mimicking abilities and tool use has caught the interest of scientists trying to understand language and intelligence.

[Page 107] In 1978, Pennsylvania psychologists David Premack and Guy Woodruff asked whether or not chimpanzees have theory of mind. Debate ensued for decades, with work by Tomasello, Hare, Call, Melis, and others ultimately suggesting chimps do have a basic ability to infer the intentions of others (Hare, 2006; Melis, 2006; Tomasello, 2005). In 2008, Call and Tomasello revisited a seminal paper, "Does the chimpanzee have a theory of mind?" and nicely reviewed the existing evidence on ape theory of mind.

CHAPTER SEVEN: A HISTORY OF VIOLENCE

[Page 121] Researchers in 2007 found early-life abuse increases the risk for juvenile arrests (Lansford), and a 2020 team reported that early-life trauma is a leading cause of later-life physical and mental illness, including mood disorders like depression, as well as an increased risk of suicidal ideation (Lippard).

CHAPTER NINE: SPEAKER WIRE

[Page 142] This number is debated, as chimp communities much larger than fifty have been reported. At one point, the group studied by John Mitani in Uganda swelled to over 200 members; yet as of early 2020, the community is in the process of splitting.

CHAPTER ELEVEN: WEATHER PERMITTING

[Page 164] *Paranthropus* evolved 2.7 million years ago, perhaps influenced by the same cooling period associated with the appearance of Homo; it was their later reliance on grasses during subsequent periods of climate change that may have led to their extinction (deMenocal, 2014).

[Page 168] In 2014, Bunn compared the prey preferences of early humans with modern lions and leopards. He analyzed antelope, wildebeest, and gazelle carcasses found at Olduvai Gorge in Tanzania, brought there and consumed by humans, most likely *Homo habilis*, nearly two million years ago. Early humans appear to have preferred hunting the adults among these large antelope species; whereas large cats will kill those of any age. When it comes to small antelopes, lions tend to go for adults in the prime life; ancestral humans only hunted older members of the herd, perhaps since hitting a smaller target with a spear required it to have been slowed by age.

CHAPTER TWELVE: A PALEOLITHIC RAW BAR

[Page 173] Isotopes are variants in chemical elements that have different numbers of neutrons. By analyzing oxygen isotopes present in deep-sea geologic samples, scientists can track Earth's temperature. Marine isotope stages show Earth underwent cyclical periods of warming and cooling. There have been five major ice ages in the history of our planet, including the Quaternary ice age, which we are still in (though the current global warming crisis may permanently change the course of our planet, we are technically still in an ice age). Throughout ice ages, the planet fluctuates between cool stretches, called glacial periods, and interglacial periods in which temperatures rise. The MIS6 cooling is the penultimate glacial period, while the most recent global cooling, the Last Glacial Period, ended around 12,000 years ago, marking the dawn of the Holocene geological epoch.

[Page 180] In 2014, the Consumer Reports National Research Center polled 1,000 Americans on their attitudes and behaviors around gluten. Over one-third of respondents reported avoiding gluten, while 63 percent believed sticking to a gluten-free diet would improve physical or mental health. The report points out the many potential downsides to avoiding gluten. Gluten-free foods often have less overall nutrients and increased levels of toxic arsenic due to the rice flour often used to replace wheat. A 2015 report found upward of 10 percent of people have a sensitivity to gluten despite not being allergic to the protein, though the pathophysiology of such cases is poorly understood (Fasano).

CHAPTER FIFTEEN: FUTURE SAPIENS

[Page 213] Chromosomes are threadlike structures composed of proteins and DNA. Humans have twenty-three pairs of them, including one pair that differs between males and females. These are the sex chromosomes, X and Y, which determine the sex of an organism. Females carry two X chromosomes and males an X and Y.

Children inherit one copy of both mom's and dad's chromosomes, and genes are shuffled around between the two, which increases genetic diversity and a species' chances of survival. The Y chromosome, however, doesn't have a partner with whom to shuffle and, with the exception of mutations, stays nearly the same through male generations. Authors of a 2018 paper showed fighting between patrilineal clans, in which family membership is determined through the father's line, could have resulted in the drop in male genetic diversity. In those patrilineal clans more successful at warfare, the Y chromosome stayed more or less the same from grandfather, through father, through sons. In clans that lost battles, it was wiped out (Zeng).

[Page 216] By looking at historical data on harvests, food prices, and local records from a small Swedish municipality called Överkalix, two reports showed that grandsons whose grandfather's had an excess of food between nine and twelve years old experience earlier mortality, including that due to diabetes. This possible epigenetic effect only appears to occur in the male lineage, with granddaughter's mortality unchanged by their grandfather's access to food (Bygren, 2001; Kaati, 2002). A 2018 study used a similar methodology as the Överkalix, looking at Swedish crop-harvest statistics from 1874 to 1910 to assess grandparents' access to food. The researchers found grandsons of grandfathers with plenty to eat more often died of cancer than those whose grandfathers weren't as well fed (Vågerö).

BIBLIOGRAPHY

Aiello LC, Wheeler P. The Expensive-tissue hypothesis: the brain and the digestive system in human and primate evolution. *Curr Anthropol.* 1995 Apr; 36(2):199–221.

Allman JM, Tetreault NA, Hakeem AY, Manaye KF, Semendeferi K, et al. The von Economo neurons in the frontoinsular and anterior cingulate cortex. *Ann N Y Acad Sci.* 2011 Apr;1225:59–71.

Almeling L, Hammerschmidt K, Sennhenn-Reulen H, Freund AM, Fischer J. Motivational shifts in aging monkeys and the origins of social selectivity. *Curr Biol.* 2016 Jul 11;26(13):1744–49.

Al-Shawaf L, Conroy-Beam D, Asao K, Buss DM. Human emotions: an evolutionary psychological perspective. *Emot Rev.* 2016 Feb 11;8(2):173–86.

Amen DG, Harris WS, Kidd PM, Meysami S, Raji CA. Quantitative erythrocyte omega-3 EPA plus DHA levels are related to higher regional cerebral blood flow on brain SPECT. *J Alzheimers Dis.* 2017;58:1189–99.

Ardila A. The evolutionary concept of "preadaptation" applied to cognitive neurosciences. *Front Neurosci.* 2016;10:103.

Arendt D, Bertucci PY, Achim K, Musser JM. Evolution of neuronal types and families. *Curr Opin Neurobiol.* 2019;56:144–52.

Arsuaga JL, Martínez I, Arnold LJ, Aranburu A, Gracia-Téllez A, et al. Neandertal roots: cranial and chronological evidence from Sima de los Huesos. *Science.* 2014 Jun 20;344(6190):1358–63.

Atkinson EG, Audesse AJ, Palacios JA, Bobo DM, Webb AE, et al. No evidence for recent selection at FOXP2 among diverse human populations. *Cell.* 2018 Sep 6; 174(6):1424–35.

Barger N, Hanson KL, Teffer K, Schenker-Ahmed NM, Semendeferi K. Evidence for evolutionary specialization in human limbic structures. *Front Hum Neurosci.* 2014;8(277):1–17.

Barger N, Stefanacci L, Schumann C, Sherwood C, Annese J, et al. Neuronal populations in the basolateral nuclei of the amygdala are differentially increased in humans compared to apes: a stereological study. *J Comp Neurol.* 2012 Sep 1; 520(13):3035–54.

Baron-Cohen S. The hyper-systemizing, assortative mating theory of autism. *Prog Neuropsychopharmacol Biol Psychiatry*. 2006 Jul;30(5):865–72.

Begley S. Amid uproar, Chinese scientist defends creating gene-edited babies. *STAT*. 2018 Nov 28. Accessed 29 February 2020.

Bennett MR, Harris JW, Richmond BG, Braun DR, Mbua E, et al. Early hominin foot morphology based on 1.5-million-year-old footprints from Ileret, Kenya. *Science*. 2009 Feb 27;323(5918):1197–201.

Benton MJ. Hyperthermal-driven mass extinctions: killing models during the Permian-Triassic mass extinction. *Philos Trans A Math Phys Eng Sci*. 2018 Oct 13;376(2130).

Benton MJ. *Vertebrate Palaeontology*. 4th ed. Hoboken, NJ: Wiley-Blackwell; 2014.

Bering JM. A critical review of the "enculturation hypothesis": the effects of human rearing on great ape social cognition. *Anim Cogn*. 2004 Oct;7(4):201–12.

Berna F, Goldberg P, Horwitz LK, Brink J, Holt S, et al. Microstratigraphic evidence of in situ fire in the Acheulean strata of Wonderwerk Cave, Northern Cape province, South Africa. *Proc Natl Acad Sci U S A*. 2012 May 15;109(20):1215–20.

Berwick RC, Chomsky N. *Why only us: Language and evolution*. Cambridge, MA: The MIT Press; 2017.

Bianchi S, Stimpson CD, Bauernfeind AL, Schapiro SJ, Baze WB, et al. Dendritic morphology of pyramidal neurons in the chimpanzee neocortex: regional specializations and comparison to humans. *Cereb Cortex*. 2013 Oct 23; 23(10):2429–36.

Boesch C. Cooperative hunting roles among Taï chimpanzees. *Hum Nat*. 2002 Mar; 13(1):27–46.

Bond M, Tejedor MF, Campbell KE Jr, Chornogubsky L, Novo N, Goin F. Eocene primates of South America and the African origins of New World monkeys. *Nature*. 2015 Apr 23;520(7548):538–41.

Bot M, Brouwer IA, Roca M, Kohls E, Penninx BWJH, et al; MooDFOOD Prevention Trial Investigators. Effect of multinutrient supplementation and food-related behavioral activation therapy on prevention of major depressive disorder among overweight or obese adults with subsyndromal depressive symptoms: the MooDFOOD randomized clinical trial. *JAMA*. 2019 Mar 5; 321:858–68.

Brain CKB, Prave AR, Hoffmann KH, Fallick AE, Botha AJ, et al. The first animals: ca. 760-million-year-old sponge-like fossils from Namibia. *S Afr J Sci*. 2012 Jan; 108(1):658.

Braun DR, Pobiner BL, Thompson JC. An experimental investigation of cut mark production and stone tool attrition. *J Arch Sci*. 2008;35:1216–23.

Brown P, Sutikna T, Morwood MJ, Soejono RP, Jatmiko, et al. A new small-bodied hominin from the late Pleistocene of Flores, Indonesia. *Nature*. 2004 Oct 28; 431(7012):1055–61.

Brunet M, Guy F, Pilbeam D, Mackaye HT, Likius A, et al. A new hominid from the Upper Miocene of Chad, Central Africa. *Nature.* 2002 Jul 11;418(6894):145–51.

Bryson V, Vogel HJ, eds. *Evolving Genes and Proteins.* Cambridge, MA: Academic Press; 1965.

Bunn HT, Gurtov AN. Prey mortality profiles indicate that Early Pleistocene *Homo* at Olduvai was an ambush predator. *Quat Int.* 2014 Feb;322–323:44–53.

Bygren LO, Kaati G, Edvinsson, S. Longevity determined by paternal ancestors' nutrition during their slow growth period. *Acta Biotheor.* 2001 Mar;49(1):53–9.

Cacioppo S, Bianchi-Demicheli F, Frum C, Pfaus JG, Lewis JW. The common neural bases between sexual desire and love: a multilevel kernel density fMRI analysis. *J Sex Med.* 2012 Apr;9(4):1048–54.

Call J, Tomasello M. Does the chimpanzee have a theory of mind? 30 years later. *Trends Cogn Sci.* 2008 May;12(5):187–92.

Calvin CM, Batty GD, Der G, Brett CE, Taylor A, et al. Childhood intelligence in relation to major causes of death in 68 year follow-up: prospective population study. *BMJ.* 2017 Jun 28;357:j2708.

Chan EKF, Timmermann A, Baldi BF, Moore AE, Lyons RJ, et al. Human origins in a southern African palaeo-wetland and first migrations. *Nature.* 2019 Nov; 575(7781):185–9.

Chang JP, Su KP, Mondelli V, Satyanarayanan SK, Yang HT, et al. High-dose eicosapentaenoic acid (EPA) improves attention and vigilance in children and adolescents with attention deficit hyperactivity disorder (ADHD) and low endogenous EPA levels. *Transl Psychiatry.* 2019;9(1):303.

Chen F, Du M, Blumberg JB, Ho Chui KK, Ruan M, et al. Association among dietary supplement use, nutrient intake, and mortality among U.S. adults: a cohort study. *Ann Intern Med.* 2019;170(9):604–13.

Chester SG, Bloch JI, Boyer DM, Clemens WA. Oldest known euarchontan tarsals and affinities of Paleocene Purgatorius to Primates. *Proc Natl Acad Sci U S A.* 2015 Feb 3;112(5):1487–92.

Consumer Reports. 6 Truths about a gluten free diet. 2014 Nov; Available at: https:// www.consumerreports.org/cro/magazine/2015/01/will-a-gluten-free-diet-really-make-you-healthier/index.htm.

Crawford MA, Sinclair AJ. The accumulation of arachidonate and docosahexaenoate in the developing rat brain. *J Neurochem.* 1972 Jul;19(7):1753–8.

D'Anastasio R, Wroe S, Tuniz C, Mancini L, Cesana DT, et al. Micro-Biomechanics of the Kebara 2 Hyoid and Its Implications for Speech in Neanderthals. *PLoS One.* 2013 Dec 18; 8(12).

Dart R. Australopithecus africanus: the Man-Ape of South Africa. *Nature.* 1925 Feb; 115(2884):195–9.

Darwin C. *The Descent of Man, and Selection in Relation to Sex.* London, England; John Murray: 1871.

Darwin C. *The Expression of the Emotions in Man and Animals*. London, England; John Murray: 1872.

DeCasien AR, Williams SA, Higham JP. Primate brain size is predicted by diet but not sociality. *Nat Ecol Evol*. 2017 Mar 27;1(5):112.

Degioanni A, Bonenfant C, Cabut S, Condemi S. Living on the edge: was demographic weakness the cause of Neanderthal demise? *PLoS One*. 2019;14(5):e0216742.

deMenocal PB. New evidence shows how human evolution was shaped by climate. *Sci Am*. 2014 Sep; Available at: https://www.scientificamerican.com/article/new-evidence-shows-how-human-evolution-was-shaped-by-climate/. Accessed 25 February 2020.

Demuth JP, De Bie T, Stajich JE, Cristianini N, Hahn MW. The evolution of mammalian gene families. *PLoS One*. 2006 Dec 20;1:e85.

Dennis MY, Nuttle X, Sudmant PH, Antonacci F, Graves TA, et al. Evolution of human-specific neural SRGAP2 genes by incomplete segmental duplication. Cell. 2012 May 11;149(4):912–22.

Derbyshire E. Brain health across the lifespan: a systematic review on the role of omega-3 fatty acid supplements. *Nutrients*. 2018 Aug 15;10(8):1094.

Détroit F, Mijares AS, Corny J, Daver G, Zanolli C, et al. A new species of Homo from the late Pleistocene of the Philippines. *Nature*. 2019 Apr;568(7751):181–6.

De Vynck JC, Anderson R, Atwater C, Cowling RM, Fisher EC, et al. Return rates from intertidal foraging from Blombos Cave to Pinnacle Point: understanding early human economies. *J Hum Evol*. 2016 Mar;92:101–15.

de Waal F. *Bonobo: The forgotten ape*. University of California Press: Berkeley, CA; 1997.

de Waal F. *The Bonobo and the Atheist*. W.W. Norton and Company: New York, NY; 2013.

di Pellegrino G, Fadiga L, Fogassi L, Gallese V, Rizzolatti G. Understanding motor events: a neurophysiological study. *Exp Brain Res*. 1992;91:176–80.

Dodd MS, Papineau D, Grenne T, Slack JF, Rittner M, et al. Evidence for early life in Earth's oldest hydrothermal vent precipitates. *Nature*. 2017 Mar 1; 543(7643):60–4.

Dunbar RI. Co-evolution of neocortex size, group size and language in humans. *Behav Brain Sci*. 1993;16(4):681–735.

Dunbar RI. Group size, vocal grooming and the origins of language. *Psychon Bull Rev*. 2017 Feb;24(1):209–12.

Dunbar RI. How conversations around campfires came to be. *Proc Natl Acad Sci U S A*. 2014 Sep 30;111(39):14013–4.

Dunbar RI. The social brain hypothesis. *Evol Anthro*. 1998;6(5):178–90.

Dunbar RI, Baron R, Frangou A, Pearce E, van Leeuwen EJ, et al. Social laughter is correlated with an elevated pain threshold. *Proc Biol Sci*. 2012 Mar 22; 279(1731):1161–7.

Dyall SC. Long-chain omega-3 fatty acids and the brain: a review of the independent and shared effects of EPA, DPA and DHA. *Front Aging Neurosci.* 2015 Apr 21; 7:52.

Dyerberg J, Bang HO, Stoffersen E, Moncada S, Vane JR. Eicosapentaenoic acid and prevention of thrombosis and atherosclerosis? *Lancet.* 1978 Jul 15; 2(8081):117–9.

Ekman P, Friesen WV. Measuring facial movement with the facial action coding system. In: Ekman P, editor. *Emotion in the Human Face.* 2nd ed. Cambridge, UK: Cambridge University Press; 2015;178–211.

Ekman P, Friesen WV, Ellsworth P. What emotion categories or dimensions can observers judge from facial behavior? In: Ekman P, editor. *Emotion in the Human Face.* 2nd ed. Cambridge, UK: Cambridge University Press; 2015;39–55.

Elston GN. Cortex, cognition and the cell: new insights into the pyramidal neuron and prefrontal function. *Cereb Cortex.* 2003;13(11):1124–38.

Enard W, Przeworski M, Fisher SE, Lai CS, Wiebe V, et al. Molecular evolution of *FOXP2*, a gene involved in speech and language. *Nature.* 2002 Aug 22; 418:869–72.

Everett D. Did Homo erectus speak? *Aoen.co.* 2018; Available at: https://aeon.co/ essays/tools-and-voyages-suggest-that-homo-erectus-invented-language.

Falk D, Zollikofer CP, Morimoto N, Ponce de León MS. Metopic suture of Taung (Australopithecus africanus) and its implications for hominin brain evolution. *PNAS.* 2012 May 29;109(22):8467–70.

Falk D, Zollikofer CPE, Ponce de León M, Smendeferi K, Alatorre Warren JL, Hopkins WD. Identification of in vivo Sulci on the External Surface of Eight Adult Chimpanzee Brains: implications for Interpreting Early Hominin Endocasts. *Brain Behav Evol.* 2018;91(1):45–58.

Fasano A, Sapone A, Zevallos V, Schuppan D. Nonceliac gluten sensitivity. *Gastroenterology.* 2015 May;148(6):1195–204.

Ferraro JV, Plummer TW, Pobiner BL, Oliver JS, Bishop LC, et al. Earliest Archaeological Evidence of Persistent Hominin Carnivory. *PLoS One.* 2013; 8(4).

Fiddes IT, Lodewijk GA, Mooring M, Bosworth CM, Ewing AD, et al. Human-Specific NOTCH2NL Genes Affect Notch Signaling and Cortical Neurogenesis. *Cell.* 2018 May 31;173(6):1356–69.

Forsyth A, Deane FP, Williams P. A lifestyle intervention for primary care patients with depression and anxiety: a randomised controlled trial. *Psychiatry Res.* 2015 Dec 15;230:537–44.

Frängsmyr T. *Linnaeus: The man and his work.* Berkeley, CA: University of California Press; 1983.

Fitch WT, de Boer B, Mathur N, Ghazanfar AA. Monkey vocal tracts are speech-ready. *Sci Adv.* 2016 Dec 9;2(12).

Fox D. What sparked the cambrian explosion? *Nature mag.* 2016 Feb 16; Available at: https://www.scientificamerican.com/article/what-sparked-the-cambrian-explosion1/. Accessed 31 January 2020.

Furuichi T. Female contributions to the peaceful nature of bonobo society. *Evol Anthropol.* 2011 Jul–Aug;20(4):131–42.

Gabbatiss J. The Monkeys That Sailed Across the Atlantic to South America. *BBC.com.* 2016 Jan 26; Available at: http://www.bbc.com/earth/story/20160126-the-monkeys-that-sailed-across-the-atlantic-to-south-america. Accessed 11 February 2020.

Galbete C, Kröger J, Jannasch F, Igbal K, Schwingshackl L, et al. Nordic diet, Mediterranean diet, and the risk of chronic diseases: the EPIC-Potsdam study. *BMC Med.* 2018 Jun 27;16(1):99.

Gazzaniga M. *Human: The science behind what makes us unique.* New York, NY: Ecco; 2008.

Genty E, Zuberbühler K. Spatial reference in a bonobo gesture. *Curr Biol.* 2014 Jul 21;24(14):1601–05.

Ghaemi N. *A first-rate madness: Uncovering the links between leadership and mental illness.* London, UK: Penguin Books; 2012.

Gilbert SL, Dobyns WB, Lahn BT. Genetic links between brain development and brain evolution. Nat Rev Genet. 2005 Jul;6(7):581–90.

Goffinet AM. The evolution of cortical development: the synapsid-diapsid divergence. *Development.* 2017 Nov 15;144(22):4061–77.

Goodall H. *In the shadow of man.* New York, NY: Collins; 1971.

Goodall J. *My friends, the wild chimpanzees.* Washington, DC: National Geographic Society; 1967.

Gómez JM, Verdú M, González-Megías A, Méndez M. The phylogenetic roots of human lethal violence. *Nature.* 2016 Oct 13;538(7624):233–7.

Gómez-Robles A, Hopkins WD, Schapiro SJ, Sherwood C. The heritability of chimpanzee and human brain asymmetry. *Proc Biol Sci.* 2016 Dec 28;283(1845).

Gonçalves B, Perra N, Vespignani A. Modeling users' activity on twitter networks: validation of dunbar's number. *PLoS One.* 2011;6(8).

Gorman J. Lab chimps are moving to sanctuaries—slowly. *The New York Times.* 2017 Nov 7; Available at: https://www.nytimes.com/2017/11/07/science/chimps-sanctuaries-research.html. Accessed 18 February 2020.

Gould SJ, Vrba ES. Exaptation-A missing term in the science of form. *Paleobiology.* 1982;8(1):4–15.

Guu TW, Mischoulon D, Sarris J, Hibbein J, McNamara RK, et al. International society for nutritional psychiatry research practice guidelines for omega-3 fatty acids in the treatment of major depressive disorder. *Psychother Psychosom.* 2019;88(5):263–73.

Haines AN, Flajnik MF, Rumfelt LL, Wourms JP. Immunoglobulins in the eggs of the nurse shark, Ginglymostoma cirratum. *Dev Comp Immunol.* 2005; 29(5):417–30.

Hambrick DZ, Tucker-Drob EM. The genetics of music accomplishment: evidence for gene-environment correlation and interaction. *Psychon Bull Rev.* 2015 Feb; 22(1):112–20.

Hamilton WD. Geometry for the selfish herd. *J Theor Biol.* 1971 May;31(2):295–311.

Hare B, Brown M, Williamson C, Tomasello M. The domestication of social cognition in dogs. *Science.* 2002 Nov 22;298(5598):1634–6.

Hare B, Call J, Tomasello M. Chimpanzees deceive a human by hiding. *Cognition.* 2006 Oct;101:495–514.

Hare B, Melis AP, Woods V, Hastings S, Wrangham R. Tolerance allows bonobos to outperform chimpanzees on a cooperative task. *Curr Biol.* 2007 Apr 3; 17(7):619–23.

Harmand S, Lewis JE, Feibel CS, Lepre CJ, Prat S, et al. 3.3-million-year-old stone tools from Lomekwi 3, West Turkana, Kenya. *Nature.* 2015 May 21; 521(7552):310–5.

Harris S. *The mystery of consciousness.* SamHarris.org. 2011.

Hatala KG, Roach NT, Ostrofsky KR. Footprints reveal direct evidence of group behavior and locomotion in Homo erectus. *Sci Rep.* 2016 Jul 12;6.

Hattori Y, Tomonaga M. Rhythmic swaying induced by sound in chimpanzees (Pan troglodytes). *PNAS.* 2019 Dec 23.

Hawks J, Wang ET, Cochran GM, Harpending HC, Moyzis RK. Recent acceleration of human adaptive evolution. *PNAS.* 2007 Dec 26;104(52):20753–8.

Hecht EE, Gutman DA, Bradley BA, Preuss TM, Stout D. Virtual dissection and comparative connectivity of the superior longitudinal fasciculus in chimpanzees and humans. *Neuroimage.* 2015 Mar;108:124–37.

Hecht EE, Gutman DA, Khreisheh N, Taylor SV, Kilner J, et al. Acquisition of Paleolithic toolmaking abilities involves structural remodeling to inferior frontoparietal regions. *Brain Struct Funct.* 2013 Sep 27;220:2315–31.

Heide M, Haffner C, Murayama A, Kurotaki Y, Shinohara H, et al. Human-specific ARHGAP11B increases size and folding of primate neocortex in the fetal marmoset. *Science.* 2020 July 30;369(6503):546–50.

Henshilwood CS, d'Errico F, van Niekerk KL, Dayet L, Queffelec A, Pollarolo L. An abstract drawing from the 73,000-year-old levels at Blombos Cave, South Africa. *Nature.* 2018 Oct;562(7725):115–8.

Herrmann E, Call J, Hernàndez-Lloreda MV, Hare B, Tomasello M. Humans have evolved specialized skills of social cognition: the cultural intelligence hypothesis. *Science.* 2007 Sep 7;317(5843):1360–6.

Herschy B, Whicher A, Camprubi E, Watson C, Dartnell L, et al. An origin-of-life reactor to simulate alkaline hydrothermal vents. *J Mol Evol*. 2014 Dec; 79(5-6):213–27.

Hill RA, Dunbar RI. Social network size in humans. *Hum Nat*. 2003 Mar; 14(1):53–72.

Hobaiter C, Byrne RW. The meanings of chimpanzee gestures. *Curr Biol*. 2014 Jul 21;24(14):1596–600.

Hoffmann DL, Standish CD, García-Diez M, Pettitt PB, Milton JA, et al. U-Th dating of carbonate crusts reveals Neandertal origin of Iberian cave art. *Science*. 2018 Feb 23;359(6378):912–5.

Hoffman HJ. The Permian extinction—when life nearly came to an end. *Nat Geo*. https://www.nationalgeographic.com/science/prehistoric-world/permian-extinction/. Accessed 20 January 2020.

Holden C. Paul MacLean and the triune brain. *Science*. 1979 Jun 8; 204(4397):1066–8.

Holloway RL, Hurst SD, Garvin HM, Schoenemann PT, Vanti WB, et al. Endocast morphology of Homo naledi from the Dinaledi Chamber, South Africa. *Proc Natl Acad Sci U S A*. 2018 May 29;115(22):5738–43.

Homer. *The Iliad*. London, UK: Penguin Classics; 1998.

Homer. *The Odyssey*. London, UK: Penguin Classics; 1999.

Houle A. Floating islands: a mode of long-distance dispersal for small and medium-sized terrestrial vertebrates. *Divers Distrib*. 1998 Jan;4(5):201–16.

Hrvoj-Mihic B, Bienvenu T, Stefanacci L, Muotri AR, Semendeferi K. Evolution, development, and plasticity of the human brain: from molecules to bones. *Front Hum Neurosci*. 2013 Oct 30;7:707.

Hublin JJ, Neubauer S, Gunz P. Brain ontogeny and life history in Pleistocene hominins. *Philos Trans R Soc Lond B Biol Sci*. 2015 Mar 5;370(1663).

Huff CD, Xing J, Rogers AR, Witherspoon D, Jorde LB. Mobile elements reveal small population size in the ancient ancestors of Homo sapiens. *Proc Natl Acad Sci U S A*. 2010 Feb 2;107(5):2147–52.

Izard CE. Emotion theory and research: highlights, unanswered questions, and emerging issues. *Annu Rev Psychol*. 2009;60:1–25.

Jacka FN. Lifestyle factors in preventing mental health disorders: an interview with Felice Jacka. *BMC Med*. 2015;13:264.

Jacka FN, O'Neil A, Opie R, Itsiopoulos C, Cotton S, et al. A randomised controlled trial of dietary improvement for adults with major depression (the 'SMILES' Trial). *BMC Med*. 2017 Jan 30;15:23.

the Jane Goodall Institute UK. Toolmaking. Available at: https://www.janegoodall.org.uk/chimpanzees/chimpanzee-central/15-chimpanzees/chimpanzee-central/19-toolmaking. Accessed 15 February 2020.

Kaas JH. Why is brain size so important: design problems and solutions as neocortex gets bigger or smaller. *Brain and Mind*. 2000;1:7–23.

Kaas JH, Balaram P. Current research on the organization and function of the visual system in primates. *Eye Brain*. 2014;6(1):1–4.

Kaati G, Bygren LO, Edvinsson S. Cardiovascular and diabetes mortality determined by nutrition during parents' and grandparents' slow growth period. *Eur. J. Hum. Genet*. 2002 Nov;10(11):682–8.

Kaati G, Bygren LO, Pembrey M, Sjöström M. Transgenerational response to nutrition, early life circumstances and longevity. *Eur. J. Hum. Genet*. 2007 Jul; 15,784–90.

Kaminski J, Bräuer J, Call J, Tomasello, M. Domestic dogs are sensitive to a human's perspective. *Behaviour*. 2009 Jul;146(7):979–98.

Kappeler PM, Watts DP. *Long-term field studies of primates*. New York, NY: Springer Publishing; 2012.

Karmin M, Saag L, Vicente M, Wilson Sayres MA, Järve M, et al. A recent bottleneck of Y chromosome diversity coincides with a global change in culture. *Genome Res*. 2015 Apr;25(4):459–66.

Kendler KS, Larsson Lönn S, Morris NA, Sundquist J, Långström N, Sundquist K. A Swedish national adoption study of criminality. *Psychol Med*. 2014 Jul; 44(9):1913–25.

Khan SU, Khan MU, Riaz H, Valavoor S, et al. Effects of nutritional supplements and dietary interventions on cardiovascular outcomes: an umbrella review and evidence map. *Ann Intern Med*. 2019 Aug 6;171(3):190–8.

King B. *How Animals Grieve*. Chicago, IL: University of Chicago Press; 2013.

King B. The Orca's Sorrow. *Sci Am*. 2019 Mar; Available at: https://www. scientificamerican.com/article/the-orcas-sorrow/.

Kniffin KM, Wilson DS. Utilities of gossip across organizational levels: multilevel selection, free-riders, and teams. *Hum Nat*. 2005 Sep;16(3):278–92.

Knoll AH, Walter MR, Narbonne GM, Christie-Blick N. The Ediacaran period: a new addition to the geologic time scale. *Lethaia*. 2007 Jan 2;39:13–30.

Koch C. Will Machines Ever Become Conscious? *Sci Am*. 2019 Dec 1. https:// www.scientificamerican.com/article/will-machines-ever-become-conscious/. Accessed 20 January 2020.

Kromhout D, Bosschieter EB, de Lezenne Coulander C. The inverse relation between fish consumption and 20-year mortality from coronary heart disease. *N Engl J Med*. 1985 May 9;312(19):1205–9.

Krupenye C, Kano F, Hirata S, Call J, Tomasello M. Great apes anticipate that other individuals will act according to false beliefs. *Science*. 2016 Oct 7; 354(6308):110–4.

Kruska D. Comparative quantitative investigations on brains of wild cavies and guinea pigs: a contribution to size changes of CNS structures due to domestication. *Mammalian Biology*. 2014;79:230–9.

Külzow N, Witte AV, Kerti L, Grittner U, Schuchardt JP, et al. Impact of Omega-3 Fatty Acid Supplementation on Memory Functions in Healthy Older Adults. *J Alzheimers Dis*. 2016;51(3):713–25.

Laland K. These amazing creative animals show why humans are the most innovative species of all. TheConversation.com. 2017; Available at: https://theconversation.com/these-amazing-creative-animals-show-why-humans-are-the-most-innovative-species-of-all-75515. Accessed 20 January 2020.

Lane N. *The Vital Question*. New York, NY: W. W. Norton & Company; 2015.

Lansford JE, Miller-Johnson S, Berlin LJ, Dodge KA, Bates JE, Pettit GS. Early physical abuse and later violent delinquency: a prospective longitudinal study. *Child Maltreat*. 2007 Aug;12(3):233–45.

Lee TH, Hoover RL, Williams JD, Sperling RI, Ravalese J, et al. Effect of dietary enrichment with eicosapentaenoic and docosahexaenoic acids on in vitro neutrophil and monocyte leukotriene generation and neutrophil function. *N Engl J Med*. 1985 May 9;9;312(19):1217–24.

Lippard ET, Nemeroff CB. The devastating clinical consequences of child abuse and neglect: increased disease vulnerability and poor treatment response in mood disorders. *Am J Psychiatry*. 2020 Jan 1;177(1):20–36.

Liu X, Somel M, Tang L, Yen Z, Jiang X, et al. Extension of cortical synaptic development distinguishes humans from chimpanzees and macaques. *Genome Res*. 2012 Apr; 22(4):611–22.

Lu ZX, Huang Q, Su B. Functional characterization of the human-specific (type II) form of kallikrein 8, a gene involved in learning and memory. *Cell Res*. 2009 Feb;19(2):259–67.

Lu ZX, Peng J, Su B. A human-specific mutation leads to the origin of a novel splice form of neuropsin (KLK8), a gene involved in learning and memory. *Hum Mutat*. 2007 Oct; 28(10):978–84.

Magadum S, Banerjee V, Murvgan P, Gangapur D, Ravikesavan R. Gene duplication as a major force in evolution. *J Genet*. 2013 Apr;92(1):155–61.

Manninen S, Tuominen L, Dunbar RI. Social laughter triggers endogenous opioid release in humans. *J Neurosci*. 2017 Jun 21;37(25):6125–31.

Maor R, Dayan T, Ferguson-Gow H, Jones KE. Temporal niche expansion in mammals from a nocturnal ancestor after dinosaur extinction. *Nat Ecol Evol*. 2017 Dec;1(12):1889–95.

Marean CW. The transition to foraging for dense and predictable resources and its impact on the evolution of modern humans. *Philos Trans R Soc Lond B Biol Sci*. 2016 Jul 5;37.

Marean CW. When the Sea Saved Humanity. *Scientific American*. 2016 Oct; Available at: https://www.scientificamerican.com/article/when-the-sea-saved-humanity1/.

Marean CW, Bar-Matthews M, Bernatchez J, Fisher E, Goldberg P, et al. Early human use of marine resources and pigment in South Africa during the Middle Pleistocene. *Nature*. 2007 Oct 18;449:905–9.

Martin D. *H. Bentley Glass, Provocative Science Theorist, Dies at 98. The New York Times*. 2005 Jan 20; Available at: https://www.nytimes.com/2005/01/20/science/h-bentley-glass-provocative-science-theorist-dies-at-98.html. Accessed 29 January 2020.

Martinac B, Saimi Y, Kung C. Ion channels in microbes. *Physiol Rev*. 2008 Oct; 88(4):1449–90.

Marques-Bonet T, Kidd JM, Ventura M, Graves TA, Cheng Z, et al. A burst of segmental duplications in the genome of the African great ape ancestor. *Nature*. 2009 Feb 12;457(7231):877–81.

Matacic C, Erard M. Can these birds explain how language first evolved? *Science*. 2018 Aug 2; Available at: https://www.sciencemag.org/news/2018/08/can-these-birds-explain-how-language-first-evolved. Accessed 20 January 2020.

McGrew WC. Savanna chimpanzees dig for food. *PNAS*. 2007 Dec 4;104(49): 19167–8.

Melis AP, Call J, Tomasello M. Chimpanzees conceal visual and auditory information from others. *J Comp. Psychol*. 2006 May;120:154–62.

Mittnik A, Massy K, Knipper C, Wittenborn F, Friedrich R, et al. Kinship-based social inequality in Bronze Age Europe. *Science*. 2019 Nov 8;366(6466):731–4.

Miyagawa S, Berwick RC, Okanoya K. The emergence of hierarchical structure in human language. *Front Psychol*. 2013 Feb 20;4:71.

Møller AP, Erritzøe J. Brain size in birds is related to traffic accidents. *R Soc Open Sci*. 2017 Mar 29;4(3):161040.

Mozzi A, Forni D, Clerici M, Pozzoli U, Mascheretti S, et al. The evolutionary history of genes involved in spoken and written language: beyond *FOXP2*. Sci Rep. 2016 Feb 25; 6:22157.

Muir J. *The Story of My Boyhood and Youth*. Boston, MA: Houghton Mifflin Company; 1913.

Natalia KG, Roach NT, Ostrofsky KR, Wunderlich RE, et al. Footprints reveal direct evidence of group behavior and locomotion in Homo erectus. *Sci Rep*. 2016;6:28766.

Nature Education. "Ion Channel." *Scitable*. 2014; Available at: https://www.nature.com/scitable/topicpage/ion-channel-14047658/. Accessed 31 January 2020.

Opie RS, O'Neil A, Jacka FN, Pizzinga J, Itsiopoulos C. A modified Mediterranean dietary intervention for adults with major depression: dietary protocol and feasibility data from the SMILES trial. *Nutr Neurosci*. 2018 Sep;21:487–501.

Palomero-Gallagher N, Zilles K. Differences in cytoarchitecture of Broca's region between human, ape and macaque brains. *Cortex.* 2019 Sep;118:132–53.

Pardo JD, Szostakiwsky M, Ahlberg PE, Anderson JS. Hidden morphological diversity among early tetrapods. *Nature.* 2017 Jun 29;546(7660):642–5.

Pargeter J, Khreisheh N, Stout D. Understanding stone tool-making skill acquisition: experimental methods and evolutionary implications. *J Hum Evol.* 2019 Aug;133:146–66.

Pascal R, Pross A, Sutherland JD. Towards an evolutionary theory of the origin of life based on kinetics and thermodynamics. *Open Biol.* 2013 Nov 6;3(11):130156.

Patel BH, Percivalle C, Ritson DJ, Duffy CD, Sutherland JD. Common origins of RNA, protein and lipid precursors in a cyanosulfidic protometabolism. *Nat Chem.* 2015 Apr 7;7(4):301–7.

Pearce E, Stringer C, Dunbar R. New insights into differences in brain organization between Neanderthals and anatomically modern humans. *Proc Biol Sci.* 2013 Mar;280(1758).

Penn JL, Deutsch C, Payne JL, Sperling EA. Temperature-dependent hypoxia explains biogeography and severity of end-Permian marine mass extinction. *Science.* 2018 Dec 7;362(6419).

Phillipson BE, Rothrock DW, Connor WE, Harris WS, Illingworth DR. Reduction of plasma lipids, lipoproteins, and apoproteins by dietary fish oils in patients with hypertriglyceridemia. *N Engl J Med.* 1985 May 9;312(19):1210–6.

Pinker S. *The better angels of our nature: Why violence has declined.* London, UK: Penguin Books; 2012.

Pinker S. *The language instinct: How the mind creates language.* New York, NY: William Morrow & Co; 1994.

Pobiner B. New actualistic data on the ecology and energetics of hominin scavenging opportunities. *J Hum Evol.* 2015 Mar;80:1–16.

Pobiner BL, Rogers MJ, Monahan CM, Harris WJK. New evidence for hominin carcass processing strategies at 1.5 Ma, Koobi Fora, Kenya. *J Hum Evol.* 2008 Jul; 55:103–30.

Portavella M, Torres B, Salas C. Avoidance response in goldfish: emotional and temporal involvement of medial and lateral telencephalic pallium. *J Neurosci.* 2004 Mar 4;24(9):2335–42.

Progovac L. Untitled review of the book *Why only us? Language and evolution. Language.* 2016;92(4):992–6.

Raghanti MA, Edler MK, Stephenson AR, Munger EL, Jacobs B, et al. A neurochemical hypothesis for the origin of hominids. *Proc Natl Acad Sci U S A.* 2018 Feb 6;115(6):E1108–E1116.

Redman LM, Smith SR, Burton JH, Martin CK, Il'yasova D, Ravussin E. Metabolic slowing and reduced oxidative damage with sustained caloric restriction support the rate of living and oxidative damage theories of aging. *Cell Metab.* 2018 Apr 3;3;27(4):805–5.e4.

Roach NT, Venkadesan M, Rainbow MJ, Lieberman DE. Elastic energy storage in the shoulder and the evolution of high-speed throwing in Homo. *Nature.* 2013 Jun 27;498(7455):483–6.

Rodríguez-Hidalgo A, Morales JI, Cebrià A, Courtenay LA, Fernández-Marchena JL, et al. The Châtelperronian Neanderthals of Cova Foradada (Calafell, Spain) used imperial eagle phalanges for symbolic purposes. *Sci Adv.* 2019 Nov 1;5(11).

Sekar A, Bialas AR, de Rivera H, Davis A, Hammond TR, et al. Schizophrenia risk from complex variation of complement component 4. *Nature.* 2016 Feb 11; 530(7589):177–83.

Seymour RS, Bosiocic V, Snelling EP, Chikezie PC, Hu Q, et al. Cerebral blood flow rates in recent great apes are greater than in Australopithecus species that had equal or larger brains. *Proc Biol Sci.* 2019 Nov 20;286(1915):20192208.

Rothman J. Daniel Dennett's Science of the Soul. *The New Yorker.* 2017 Mar 20; Available at: https://www.newyorker.com/magazine/2017/03/27/daniel-dennetts-science-of-the-soul. Accessed 20 January 2020.

Sarich VM, Wilson AC. Immunological time scale for hominid evolution. *Science.* 1967 Dec 1;158(3805):1200–3.

Sayol F, Maspons J, Lapiedra O, Iwaniuk AN, Székely T, Sol D. Environmental variation and the evolution of large brains in birds. *Nature Commun.* 2016 Dec 22;7(13971).

Schirrmeister BE, Gugger M, Donoghue PC. Cyanobacteria and the great oxidation event: evidence from genes and fossils. *Palaeontology.* 2015 Sep;58(5):769–85.

Schmelz M, Grueneisen S, Kabalak A, Jost J, Tomasello M. Chimpanzees return favors at a personal cost. *Proc Natl Acad Sci U S A.* 2017 Jul 11;114(28):7462–67.

Semendeferi K, Armstrong E, Schleicher A, Zilles K, Van Hoesen GW. Limbic frontal cortex in hominoids: a comparative study of area 13. *Am J Phys Anthropol.* 1998 Jun;106:129–55.

Semendeferi K., Damasio H, Frank R, Van Hoesen GW. The evolution of the frontal lobes: a volumetric analysis based on three-dimensional reconstructions of magnetic resonance scans of human and ape brains. *Journal of Human Evolution.* 1997 Apr;32(4):375–88.

Semendeferi K, Lu A, Schenker N, Damasio H. Humans and great apes share a large frontal cortex. *Nat Neurosci.* 2002 Mar;5(3):272–6.

Semendeferi K, Schleicher A, Zilles K, Armstrong E, Van Hoesen, GW. Prefrontal cortex in humans and apes: a comparative study of area 10. *American Journal of Physical Anthropology.* 2001;114(3):224–41.

Semendeferi K, Teffer K, Buxhoeveden DP, Park MS, Bludau S, et al. Spatial organization of neurons in the prefrontal cortex sets humans apart from great apes. *Cereb Cortex*. 2011 Jul;21:1485–97.

Schenker NM, Buxhoeveden DP, Blackmon WL, Amunts K, Zilles K, Semendeferi K. A comparative quantitative analysis of cytoarchitecture and minicolumnar organization in Broca's area in humans and great apes. *J Comp Neurol*. 2008 Sep 1;510(1):117–28.

Shen J, Chen J, Algeo TJ, Yuan S, Feng Q, et al. Evidence for a prolonged Permian-Triassic extinction interval from global marine mercury records. *Nat Commun*. 2019 Apr 5;10(1):1563.

Shennan S, Downey S, Timpson A, Edinborough K, Colledge S, et al. Regional population collapse followed initial agriculture booms in mid-Holocene Europe. *Nat Commun*. 2013 Oct 1;4(2486).

Shubin N. *Your inner fish: A journey into the 3.5-billion-year history of the human body*. New York, NY: Pantheon; 2008.

Shultz S, Nelson E, Dunbar RIM. Hominin cognitive evolution: identifying patterns and processes in the fossil and archaeological record. *Philos Trans R Soc Lond B Biol Sci*. 2012 Aug 5;367(1599):2130–40.

Sliwa J, Freiwald WA. A dedicated network for social interaction processing in the primate brain. *Science*. 2017 May 19:356(6339):745–9.

Smith EI, Jacobs Z, Johnsen R, Ren M, Fisher EC, et al. Humans thrived in South Africa through the Toba eruption about 74,000 years ago. *Nature*. 2018 Mar 22; 555(7697):511–5.

Sniekers S, Stringer S, Watanabe K, Jansen PR, Coleman JRI, et al. Genome-wide association meta-analysis of 78,308 individuals identifies new loci and genes influencing human intelligence. *Nat Genet*. 2017 Jul;49(7):1107–12.

Steele EJ, Al-Mufti S, Augustyn KA, Chandrajith R, Coghlan SG, et al. Cause of Cambrian Explosion—Terrestrial or Cosmic? *Prog Biophys Mol Biol*. 2018 Aug; 136:3–23.

Stephenson-Jones M, Samuelsson E, Ericsson J, Robertson B, Grillner S. Evolutionary conservation of the basal ganglia as a common vertebrate mechanism for action selection. *Curr Biol*. 2011 Jul 12;21(13):1081–91.

Stetka B. The Best Diet for Your Brain. *Sci Am*. 2016 Mar; Available at: https://www.scientificamerican.com/article/the-best-diet-for-your-brain/.

Stetka B. Cocktail of brain chemicals may be a key to what makes us human. *Sci Am*. 2018 Jan 24; Available at: https://www.scientificamerican.com/article/cocktail-of-brain-chemicals-may-be-a-key-to-what-makes-us-human/. Accessed 20 January 2020.

Stetka B. Food for thought: do we owe our large primate brains to a passion for fruit? *Sci Am.* 2017 Mar 27; Available at: https://www.scientificamerican.com/article/food-for-thought-do-we-owe-our-large-primate-brains-to-a-passion-for-fruit/. Accessed 20 January 2020.

Stetka B. Lab-grown "mini brains" can now mimic the neural activity of a preterm infant. *Sci Am.* 2019 Jan 24; Available at: https://www.scientificamerican.com/article/lab-grown-mini-brains-can-now-mimic-the-neural-activity-of-a-preterm-infant/. Accessed 20 January 2020.

Stetka B. Monkeys have a specialized brain network for sizing up others' actions. *Sci Am.* 2017 May 18; Available at: https://www.scientificamerican.com/article/monkeys-have-a-specialized-brain-network-for-sizing-up-others-rsquo-actions/. Accessed 20 January 2020.

Stetka B. Steven Pinker: This is history's most peaceful time—New study: "Not so fast." *Sci Am.* 2017 Nov 9; Available at: https://www.scientificamerican.com/article/steven-pinker-this-is-historys-most-peaceful-time-new-study-not-so-fast/. Accessed 20 January 2020.

Stevens NJ, Seiffert ER, O'Connor PM, Roberts EM, Schmitz MD, et al. Palaeontological evidence for an Oligocene divergence between Old World monkeys and apes. *Nature.* 2013 May 15;497:611–4.

Stout D, Hecht EE. Evolutionary neuroscience of cumulative culture. *PNAS.* 2017 Jul 25;114(30):7861–8.

Stout D, Hecht EE, Khreisheh N, Bradley B, Chaminade T. Cognitive demands of lower Paleolithic toolmaking. *Plos One.* 2015 Apr 15; 10:e0121804.

Surbeck M, Boesch C, Crockford C, Thompson ME, Furuichi T, et al. Males with a mother living in their group have higher paternity success in bonobos but not chimpanzees. *Curr Biol.* 2019 May 20;29(10).

Suzuki IK, Gacquer D, Van Heurck R, Kumar D, Wojno M, et al. Human-specific *NOTCH2NL* genes expand cortical neurogenesis through delta/notch regulation. *Cell.* 2018 May 31;31;173(6):1370–84.

Takahashi K, Yamanaka S. Induction of pluripotent stem cells from mouse embryonic and adult fibroblast cultures by defined factors. *Cell.* 2006 Aug 25; 126(4):663–76.

Tan J, Hare B. Bonobos share with strangers. *PLoS One.* 2013;8(1):e51922.

Teffer D, Buxhoeveden D, Stimpson CD, Fobbs AJ, Schapiro SJ, et al. Developmental changes in the spatial organization of neurons in the neocortex of humans and chimpanzees. *J Comp Neurol.* 2013;521:4249–59.

Tian R, Gachechiladze MA, Ludwig CH, Laurie MT, Hong JY, et al. CRISPR interference-based platform for multimodal genetic screens in human iPSC-derived neurons. *Neuron.* 2019 Oct 23;104(2):239–55.e12.

Tiihonen J, Rautiainen MR, Ollila HM, Repo-Tilhonen E, Virkkunen M, et al. Genetic background of extreme violent behavior. *Mol Psychiatry*. 2015 Jun; 20(6):786–92.

Tomasello M. The ontogeny of cultural learning. *Curr Opin Psychol*. 2016 Apr;8:1–4.

Tomasello M, Call J. The role of humans in the cognitive development of apes revisited. *Anim Cogn*. 2004 Oct;7(4):213–5.

Tomasello M, Carpenter M, Call J, Behne T, Mall H. Understanding and sharing intentions: the origins of cultural cognition. *Behav Brain Sci*. 2005 Oct;28: 675–91.

Tomer R, Denes A, Tessmar-Raible K, Arendt D. Profiling by image registration reveals common origin of annelid mushroom bodies and vertebrate pallium. *Cell*. 2010 Sep 3;142(5):800–9.

Trinkaus E, Samsel M, Villotte S. External auditory exostoses among western Eurasian late Middle and late Pleistocene humans. *PLoS ONE*. 2019 Aug 14;14(8).

Trujillo CA, Gao R, Negraes PD, Gu J, Buchanan J, et al. Complex oscillatory waves emerging from cortical organoids model early human brain network development. *Cell Stem Cell*. 2019 Oct 3;25(4):558–69.e7.

Turkheimer E. Three laws of behavior genetics and what they mean. *Curr Dir Psychol Sci*. 2000;9:160–4.

Vågerö D, Pinger PR, Aronsson V, van den Berg GJ. Paternal grandfather's access to food predicts all-cause and cancer mortality in grandsons. *Nat Commun*. 2018;11;9(1):5124.

Vaidyanathan G. How have hominids adapted to past climate change? *Sci Am*. 2010 Apr 13; Available at: https://www.scientificamerican.com/article/hominids-adapt-to-past-climate-change/.

Wang ET, Kodama G, Baldi P, Moyzis RK. Global landscape of recent inferred Darwinian selection for Homo sapiens. *PNAS*. 2006 Jan 3;103(1):135–40.

Warneken F, Rosati AG. Cognitive capacities for cooking in chimpanzees. *Proc Biol Sci*. 2015 Jun 22;282(1809).

Washburn S. *Ape Into Man; A Study of Human Evolution*. Boston, MA: Little, Brown; 1973.

Wicht H, Northcutt RG. Telencephalic connections in the Pacific hagfish (Eptatretus stouti), with special reference to the thalamopallial system. *J Comp Neurol*. 1998 Jun 1;395(2):245–60.

Wiessner PW. Embers of society: firelight talk among the Ju/'hoansi Bushmen. *Proc Natl Acad Sci U S A*. 2014 Sep 30;111(39):14027–35.

Wilfred J. *Integrative action of the autonomic nervous system: Neurobiology of homeostasis*. Cambridge, UK: Cambridge University Press; 2008.

Wilson ML, Boesch C, Fruth B, Furuichi T, Gilby IC, et al. Lethal aggression in Pan is better explained by adaptive strategies than human impacts. *Nature.* 2014 Sep 18;513:414–7.

Winslow JT, Insel TR. The social deficits of the oxytocin knockout mouse. *Neuropeptides.* 2002;36(2–3):221–9.

Wobber V, Hare B, Wrangham R. Great apes prefer cooked food. *J Hum Evol.* 2008 Aug;55(2):340–8.

Wong E, Mölter J, Anggono V, Degnan SM, Degnan BM. Co-expression of synaptic genes in the sponge Amphimedon queenslandica uncovers ancient neural submodules. *Sci Rep.* 2019 Oct 31;9(1):15781.

Wong, K. Ancient Cave Paintings Clinch the Case for Neandertal Symbolism. *Sci Am.* 2018 Feb 23; Available at: https://www.scientificamerican.com/article/ancient-cave-paintings-clinch-the-case-for-neandertal-symbolism1/.

Wrangham RW. *Catching fire: how cooking made us human.* New York, NY: Basic Books; 2009.

Wrangham RW. *The goodness paradox: the strange relationship between virtue and violence in human evolution.* New York, NY: Pantheon; 2019.

Wrangham RW, Peterson D. *Demonic males: apes and the origins of human violence.* Boston, MA: Houghton Mifflin Harcourt; 1996.

Xu K, Schadt EE, Pollard KS, Roussos P, Dudley JT. Genomic and network patterns of schizophrenia genetic variation in human evolutionary accelerated regions. *Mol Biol Evol.* 2015 May;32(5):1148–60.

Yang X, Dunham Y. Minimal but meaningful: probing the limits of randomly assigned social identities. *J Exp Child Psychol.* 2019 Sep;185:19–34.

Yerkes R. *Almost Human.* New York, NY: The Century Co; 1925.

Zanella M, Vitriolo A, Andirko A, Martins PT, Sturm S, et al. Dosage analysis of the 7q11.23 Williams region identifies *BAZ1B* as a major human gene patterning the modern human face and underlying self-domestication. *Sci Adv.* 2019 Dec 04;5(12).

Zeng TC, Aw AJ, Feldman MW. Cultural hitchhiking and competition between patrilineal kin groups explain the post-Neolithic Y-chromosome bottleneck. *Nat Commun.* 2018 May 25;9(1):2077.

Zhou CF, Wu S, Martin T, Luo ZX. A Jurassic mammaliaform and the earliest mammalian evolutionary adaptation. *Nature.* 2013 Aug 8;500(7461):163–7.

Zimmer C. The Planet Has Seen Sudden Warming Before. It Wiped Out Almost Everything. *The New York Times.* 2018 Dec 7; Available at: https://www.nytimes.com/2018/12/07/science/climate-change-mass-extinction.html. Accessed 11 February 2020.

PHOTO AND ILLUSTRATION CREDITS

All illustrations, except noted below, by Tim Phelps

INDEX

abiogenesis, 27–28
acetylcholine, 157–158
Acheulean tools, 188
adrenaline, 145
aggressive behavior. *See also* violence
 acetylcholine levels, 156–157
 cultural pressures in suppressing, 122
 domestication process in controlling,
  152, 154, 156, 213
 genetic influences, 121–122
 reactive and proactive, 152
agriculture, effects of, 201–204
Aiello, Leslie, 168–169
Akawaio people, 113
Alexa, 224, 225, 226
algae
 branching off, 32
 phytosynthesis and oxygen levels,
  41–42, 56–57
 as source of healthy fats, 176, 177
Almost Human (Yerkes), 125
altruism, 104–105
Alzheimer's disease, 176, 182, 183, 207
American Museum of Natural History,
 74–75
amniotes, 46
amygdala
 early bony fish, 42
 function, 49, 145–146
 human fear and, 42, 145, 147
 neuronal composition, 92–93
 paleomammalian brain, 48
 in wild animals, 154

animism, 197
anthropoids, 58
apes
 ape encounters, 113–116
 brain blood flow rate, 78
 Center For Great Apes, 23
 communication, 135–136
 distinguishing traits, 66
 in Eurasia, 66–67
 food sources, 165
 human-like behaviors, 21
 oldest known, 66
 in popular culture, 114–115
 selection for terrestrial living, 70–71
 split from chimpanzees, 69, 71
 spread of, 66–67
 state of research, 23, 130
ape sanctuaries, 23
arboreal life, 58–59, 164–165
archaea, 30
Ardipithecus (Ardi), 72
Arendt, Detlev, 37, 234n
art
 cave, 85, 194–195, 235n
 music and rhythm, 194–196
 oldest known works, 194–195
 storytelling, 194
 visual, 194–196
artificial intelligence
 defining consciousness, 223–226
 difficulty of mimicking biological
  functions, 226
 future of, 226–227

(IIT) integrated information theory, 224, 225
  in popular culture, 224
  virtual personal assistants (Alexa and Siri), 225
artisans, 194
asteroid impact, 56
asymmetry, 140–141
Australopithecines
  brain, 78–80
  as colloquial name for *Australopithecus*, 73
  Lucy, 74, 187
  physical characteristics, 164
  "robust," 76
  use of tools, 187
Australopithecus
  brain blood flow rate, 78
  brain evolution, 78–80
  colloquial names for, 73
  crude butchering by, 167
  physical description, 86
  transition to *Homo*, 80, 89
Australopithecus afarensis, 76, 83, 163
Australopithecus africanus, 72–73, 75–76, 79, 83
Australopithecus anamensis, 75
Australopithecus garhi, 80
Australopiths
  appearance and traits, 74
  brain size, 83
  as colloquial name for *Australopithecus*, 74
  diorama, 73–74
  evidence of upright walking, 75
  toolmaking, 78
  vocabulary, 144
autism, 214–215

bacteria, 14, 30, 31
Baggini, Julian, 143
Baron-Cohen, Simon, 214
Battel, Andrew, 114
Begun, David, 66, 67, 71, 77

behavioral genetics, laws of, 137–138
beliefs, shared, 198–199
Belyaev, Dmitry, 154–155
Berger, Lee, 82
Berwick, Robert C., 137
bilateral animals, 34, 37, 39, 41, 234n
Bilateria / Bilateria, 34, 36, 39–41
bipedalism, advantages of, 76–77
  See also upright posture
bipolar disorder, 208
bird feathers, 138
birds
  brain size and sociality, 236n
  crows, 185, 186, 236n
  domesticated species, 155–156
  species that live in changing environments, 186
birdsong, 137
birth
  babies' heads at, 90
  obstetrical dilemma, 90–91, 148, 172
  synaptic pruning at, 208
blood flow, brain, 78, 109, 178–179
Blue Zone regions, 181
body temperature, 58–59
Boesch, Christophe, 186
bone formation, 41–42, 235n
bone marrow in diet, 168
The Bonobo and the Atheist (de Waal), 100
Bonobo Conservation Initiative, 130
bonobos
  agreeable personalities, 18, 19–20, 127
  appearance, 18
  Belle, 18, 23
  chimpanzees compared to, 18, 19–20, 100, 127, 129–130, 131–133, 136, 156
  compassion, 100
  folklore, 130
  human-like behaviors, 19–20, 21
  maternal intervention in sons' sexual behavior, 130–131, 132
  matriarchal societies, 20
  moral behavior, 128
  sexual behavior, 20, 127–128

Boule, Marcellin, 84–85
brain architecture. *See also* amygdala;
    prefrontal cortex
    anterior cingulate cortex, 92, 110
    ARHGAP11B gene, 235n
    brainstem, 40, 42, 47–49, 149
    Broca's area, 140, 141, 144
    Brodmann's Area 10, 93–94
    cerebellum, 42, 47–48
    cerebral cortex, 47, 48, 89, 92
    cranial characteristics of common
        ancestor, 91–92
    folds (sulci), 62, 63, 79, 80, 188
    foramen magnum, 73
    frontal lobe, 47, 63, 90, 140
    insular cortex, 92
    language center, 89, 140, 141, 144
    motor cortex, 62, 63
    occipital lobe, 47, 48, 79
    parietal lobe, 47, 48, 63, 188, 189
    plumbing and wiring analogy, 50
    sensory cortex, 62, 63
    speech center, 140, 141
    supramarginal gyrus, 188
    temporal lobe, 47, 48, 140
    use of term, 62
    Wernicke's area, 140, 141, 144
brain size
    anatomical evolution and, 101
    big-brained hominins, 87
    birds, 236n
    changes over time, 83–84
    dietary preferences and, 102–103
    exploitation of the environment and,
        102
    genetic contributions to, 206–207
    obstetrical dilemma, 90–91, 148, 172
    Sapien / Neanderthal comparisons,
        85–86, 91
    simian species, 62–63
    skulls, 52, 78, 90–91
    social groups and, 101, 236n
brain size, measuring. *See* encephaliza-
    tion quotients

Broca's area, 140, 141, 144
Brodmann's Area 10, 93–94
Brunet, Michel, 71
Budongo Forest Reserve (Uganda), 135
Buettner, Dan, 181
Bunn, Henry, 168, 237n
Buss, David, 151

calories
    bulbs, tubers, and rhizomes
        (geophytes), 165, 173–174
    fruits, seeds, and flowers, 164–165
    importance of, 179
    intake as prime concern, 184
    meat, 166–167, 169
    sugar, 179
Cambrian explosion, 34
Cambrian life, 40
camps, 103–107, 143–144
carnivorism, 12, 166, 167–168
Catching Fire: How Cooking Made Us
    Human (Wrangham), 169, 172
cave art, 85, 194–195, 235n
cemeteries, 197
Center For Great Apes, 23
cephalization, 40
Chalmers, David, 223
childhood, prolonged, 51, 91
Chim and Panzee, 125–126
chimpanzees
    Amos, 100
    appearance, 18
    bonobos compared to, 18, 19–20, 100,
        127, 129–130, 131–133, 136, 156
    border patrolling, 116–117
    brain size, 52
    cohabitation with humans, 152–153
    communication, 135–136, 144
    creative behavior, 185–186
    death of Thomas, 150
    encephalization quotient, 63
    frontal pyramidal branching, 92
    Godi, 115
    human-like behaviors, 19–21

hunting behavior, 165, 185–186

inferring the intentions of others, 107

killing behavior, 20, 115–119

language and group size, 142, 236

observations by Goodall, 20–21, 115, 185, 189, 233n

reciprocal altruism, 105

Travis, 153

violent behavior, 20, 115–119, 121, 122–123, 153

choanoflagellates, 32–33

Chomsky, Noam, 137, 139

chordates, 40–41

chromosomes, 13, 213, 237–238n

churches / places of worship, 197

Civilized to Death (Ryan), 119

climate change

effects on food supply, 25, 103, 163–164, 236n

extinction of ape species, 67

migrations, 66–67, 83

warming, 55, 67

coastal living

Pinnacle Point (South African coast), 173–174, 193

seafood, 174, 175–177

collaboration, creative, 190, 196–199

Collins, Francis, 181, 220

color, 192–193, 196

comb jellies, 34, 35, 37, 39

common ancestor

in evolution of central nervous system, 234n

mitochondrial Eve, 65, 82

preadapted to evolve symbolic speech, 138

Sahelanthropus tchadensis (Toumaï), 71–72

sponges as, 26, 32

communal living. *See also* social behavior

benefits of, 105–106

camps, 103–107, 143–144

cooperation, 99–100, 107, 164, 198–199

finding food, 62

group membership, 106–107

group selection, 104–105

as predictor of happiness, 216

safety, 61, 105

communication

early humans, 138

gesture, 135–136, 138, 142, 144, 153

gossip, 143, 217

neurocircuitry of, 140, 144

social brain hypothesis and, 141–142

symbolic language, 99, 140, 144, 159

virtual, 216–218

consciousness, 27, 223–226

convergent evolution, 156–157

Coolidge, Harold, 125

cooperation, 99–100, 107, 164, 198–199

cortisol, 145, 151

COVID-19 pandemic, 212

cows, 150, 151

Crawford, Michael, 176, 177

The Creative Spark (Fuentes), 187

Cretaceous–Paleogene extinction, 56–57

Cretaceous period, 47, 56

CRISPR-Cas9, 219–221

crops, 201

"cuddle hormone," 111, 149

cultural learning as human social learning, 202

Cunnane, Stephen, 178, 216

cyanobacteria, 31

Dart, Raymond, 72–73, 78

Darwin, Charles

on affection and grief, 150

on the biological basis of emotions, 146

The Descent of Man, 65, 99, 114, 137, 143

descent with modification, 29

on domesticated animals, 153–154

The Expression of the Emotions in Man and Animals, 146

on fire and cooking food, 170

on functions of traits through generations, 138

on jaws and teeth, 166
on kinship among primates, 65, 114
on language, 137, 143
on social instincts, 99
work with Wallace, 13, 29, 233n
Darwin awards, 213
Darwinian evolution, 13, 29
DASH diet, 183
dating apps, 218
Dawkins, Richard, 104–105, 226
DeCasien, Alexandra, 61–62, 102–103
deMenocal, Peter B., 163
dementia, 176, 181
Demonic Males: Apes and the Origins of Human Violence (Wrangham), 118
Denisovans
    extinction of, 94
    language acquisition gene variants, 139
    shared DNA, 87, 89, 128
Dennet, Daniel, 226
depression, 177, 183, 208, 226–227, 234n, 236n
Descartes, René, 223, 225
The Descent of Man (Darwin), 65, 99, 114, 137, 143
Devonian period, 43
De Vynck, Jan, 174
de Waal, Frans, 100, 127, 128, 133
DHA (docosahexaenoic acid), 176, 177, 178
Diamond, Jared, 202
diet
    Alzheimer's disease and, 176, 182, 183, 207
    brain health and, 170, 176–177, 181–184
    correlational research, 180, 184
    dementia and, 176, 181
    eating plans, 183–184
    fish-fixated regions, 183
    food patterns, 181–184
    healthy, 171, 181, 183–184
    high in sugar, 183
    hippocampus size and, 183
    interpreting nutritional studies, 180–181

interpreting research, 180
    omega-3s, 176–178, 180–181
    during pregnancy, 183
    raw diet, 170–171
dinosaurs, 46, 47, 56
DNA
    Denisovan, 87, 89, 128
    epigenetic changes, 215–218
    gene-editing technologies, 219–220, 219–221
    Neanderthal, 87, 128
    origins, 27–28
    propagation of, 104
    shared, 22, 128, 133
docosahexaenoic acid (DHA), 176, 177
dogs, 64, 153
dolphins, 64, 185
domesticated environments, 155–156
domestication. *See also* self-domestication
    of animals, 152–155
    in control of aggressive behavior, 152, 154, 156, 213
    distinguished from tameness, 152–153
    evolutionary byproducts, 155–157
domestication syndrome, 154, 156, 157
Donn, Paul (ape), 17–18
dopamine, 35, 122, 149, 157, 158, 234n
droughts, 165, 173, 202
Dryopithecus / dryopithecines, 66–67
dualism, 223
Dudley, Joel, 208–209
Dunbar, Robin, 101–102, 103, 142–143, 189, 196

early-life abuse, 121, 236n
ecological hypothesis of intelligence, 102–103
Ediacaran period, 32, 33
eicosapentaenoic acid (EPA), 176
Ekembo / ekembos, 66
Ekman, Paul, 147
The Elementary Particles (Houellebecq), 229
elephants, 92, 151

elephants**PM also dogs and dolphins, 92, 151
emergence, concept of, 225–226
emotions
  adaptive function, 151–152
  amygdala processing of, 145–146
  animal / human parallels, 146
  biological basis for, 146–147
  evolution of, 147–148
  rational thought vs., 157
  research on, 148
empathy
  development of, 108, 120
  in effective leadership, 208
  neuronal aspects of, 110–111
  selection for, 158
encephalization quotients
  problematic aspects, 63
  relative values, 62–63, 64, 84
  use of, 52
endocasts, 62, 70, 79–80, 86, 88, 91
endorphins, 99, 142, 196
endosymbiosis, 233n
energy. See also calories
  brain as energy-expensive, 53, 171
  cooked food as efficient source of, 169–171
  delivery and blood flow, 78
epigenetics, 13, 215–218
epinephrine (adrenaline), 145. See also norepinephrine
Eskimo cardiovascular health, 176
estrus females, 127, 128
eugenics movement, 205
eukaryotes, 31, 32
eusociality, 103–104
Eve hypothesis, 65
Everett, Daniel, 192
evolution
  biological, 204, 215
  community-based, 108–109
  convergent, 156–157
  cultural, 201–204, 215
  Darwinian, 29, 34
  divergences, 64, 65, 235n
  family tree, 83
  genetic drift, 29
  regional micro-evolution, 214–215
evolutionary byproducts, 187–188, 196, 226
exaptations. See preadaptations
The Expression of the Emotions in Man and Animals (Darwin), 146
extinctions
  Cretaceous–Paleogene, 56–57
  of human species, 94
  large animal species, 227
  megafauna, 227
  MIS6 near-extinction, 173, 174, 237n
  near-extinctions, 175
  Permian, 47, 55

F5 brain region, 140
facial expressions, 147
facial features, 157
Falk, Dean
  on effects of modern technology, 217
  on food sources and brain evolution, 179
  on language, 136
  on musicality and language, 196
  skull study, 90–91
  on sulcal patterns, 80
false beliefs, 107–108
fat and brain composition, 176
fear, 145, 146, 147
feet, 86
Feldman, Marcus, 213
females / women
  bonobo intervention in sons' sexual behavior, 130–131, 132
  chimpanzees vs. bonobos, 127
  cultural cost of cooking, 172
  dominance, 20, 127
  male and female sizes, 75, 86
  matrilineal human groups, 202
  mitochondrial Eve hypothesis, 65, 82
  obstetrical dilemma, 90–91

as property, 203
roles in farming societies, 202
violence toward, 122–123
fight or flight response, 49, 145
fingers and hands, 86
fire
 cooked food, 169–171
 cultural cost for females, 172
 social aspects, 143–144, 172
fire and cooking
fish in evolutionary course, 42–46
fish oils, 176
folds (sulci), 62, 63, 79, 188
fontanelles, 90
food. *See also* calories; diet
 aquatically sourced nutrients, 178
 climate change effects on supply, 25,
  103, 163–164, 236n
 competition for, 107
 cooked, 169–171
 marine, 176–177
 mollusks, 174
 preserving, 171
 raw food movement, 170–171
 scarcity, 130
 seafood, 174, 175–177
food sourcing
 advantage of groups, 62
 foraging, 102, 174, 175, 178, 179, 193
 hunting, 185–186
 prey preferences, 168, 237n
 scavenging, 167, 168
footprints, 75, 77, 86
Foradada Cave site, 95
foraging, 102, 119, 174, 175, 178, 179
fossils
 documenting human / ape split, 71–72
 evidence of animal origin, 32, 33
 first vertebrate life-form on land, 44
 foot bones, 58
 limitations of knowledge from, 88
 placental mammal, 44
 records of dinosaurs and marine
  reptiles, 57

synapsid, 52
foxes, 154–155
Freiwald, Winrich, 109–110
fruit-eating species (frugivores), 102
Fuentes, Agustín, 187, 190, 196

Galton, Francis, 205
Gazzaniga, Michael, 107, 109
gene duplications, 22, 88–89, 233n
gene-editing technologies, 219–221
genes. *See also* DNA
 aggressive behavior and, 121–122
 ARHGAP11B, 235n
 ASPM, 206
 BAZ1B, 157
 C4 (synaptic pruning), 207
 CDH13 (or cadherin 13), 122
 craniofacial, 157
 FOXP2, 139
 heritable behavioral traits, 137–138
 L0, 82–83
 MAOA ("warrior gene"), 122
 nature vs. nurture, 137–138, 205–206
 NOTCH2NL, 89
 selfish, 105
 SRGAP2C, 89
gene-sequencing technology, 64–65
genetic diversity, 175, 203, 212–213,
 237–238n
genetic engineering, 13, 218–222
genetic mutations
 contributing to brain size, 206
 diseases caused by, 219
 natural selection on, 29, 138, 193
germline editing, 219–220
gesture
 animal understanding of, 153
 chimpanzees, 144
 early humans, 138
 expressing displeasure, 135–136
 overlap between chimps and
  bonobos, 136
 vocal grooming, 142
Ghaemi, Nassir, 208

glacial period, 173, 237n
Glass, H. Bentley, 228
gluten, avoiding, 237n
gods, 197
Gombe National Park (Tanzania), 20, 115, 118–119
Gómez, José María, 120–121
Goodall, Jane, 20–21, 115, 185, 189, 233n
The Goodness Paradox (Wrangham), 151
gorillas
    branching off, 65
    encephalization quotient, 64
    human encounters with, 113–114
    Paul Donn, 17–18
    shared anatomy and characteristics, 114
    skeletons, 70
Gould, Stephen Jay, 138
Graham, Kirsty, 136
Great Ape Dictionary, 135
Great Dying, 55
Great Oxidation Event, 31
grief, 149–151
The Grisly Folk (Wells), 85
grooming, 99–101
grooming networks, 102

Hadean eon, 27, 28
hagfish, 42
Hambrick, David Z., 205, 206
hands and fingers, 77, 86
Hanno the Navigator, 113–114
haplorhines ("dry-nosed" primates), 58, 59
happiness, predictor of, 217
Harari, Yuval Noah, 194, 197–198, 204
Hare, Brian
    on bonobo behavior, 20, 129–130
    chimpanzees' inference of intentions of others (study), 107, 236n
    on domesticated animals and human social cues, 153, 155
    on friendliness as survival strategy, 154
    on shared behaviors, 133
    Survival of the Friendliest, 154

Hare, Vanessa, 129
Harris, Sam, 223, 225–226
head, evolution of, 40
hearing, 52–53, 56, 60
He Jiankui, 219–220
Her (movie), 224
Herrman, Esther, 109
Hill, Andrew, 75
hirsuteness, 77
Hobaiter, Catherine, 102, 135–136
Hobbes, Thomas, 115
Holloway, Ralph
    on brain anatomy and circuity, 86, 94
    on brain size and intelligence, 78–79, 86
    on human trajectory, 71
    lab, 69–70
    on understanding our cognitive past, 88, 89
hominids, 18, 64
Hominin / hominins
    defined, 64
    endocasts of fossil record, 70
    family tree, 64–65
    first upright bipedal, 72–73
    fossils, 72
    teeth, 86
Homo
    acquisition of language, 144
    branching off, 80
    as new genus, 163
    pace of evolutionary change, 101
Homo denisova, 81. *See also* Denisovans
Homo erectus
    adaptability, 163
    brain size, 83
    branching off, 81
    campsites, 104
    communication, 144
    diet, 168
    encephalization quotient, 84
    migration, 144, 164
    mouth, 166
    throwing, 86–87
    toolmaking, 187, 191–192
    use of fire, 171

Homo ergaster, 81
Homo floresiensis, 81, 83, 84
Homo habilis
   brain size, 83
   branching off, 81
   diet, 167, 179, 237n
   encephalization quotient, 84
   language, 144
   mouth, 166
   toolmaking, 187
Homo heidelbergensis, 81, 83, 84
Homo luzonensis, 82
Homo naledi, 82, 91
Homo neanderthalensis, 81, 84. *See also*
   Neanderthals
Homo sapiens
   damage inflicted by, 227–228
   domestication process, 154
   domination by, 94
   endurance, 13, 25
   migration, 201, 227
   name, 80
   throwing, 86–87
Houellebecq, Michel, 229
humans
   biological relatives, 19, 22
   encephalization quotient, 63
   schism with other apes, 71
   small-bodied, 81–82
hunter-gatherers, 119, 179
hunting
   chimpanzee hunting behavior, 165,
   185–186
   human stamina in, 87
   prey preferences, 237n
   as a social activity, 164
Huxley, Thomas, 114
hydrogen cyanide, 27–28
hyoid bone, 137

ice ages, 237n
imaging technologies, 49, 63, 109, 110,
   188–189
induced pluripotent stem cells (iPSCs),
   220

information sharing, 101, 103, 138,
   190–191, 201, 228
innovation
   creative intelligence, 187–188,
   189–192
   cultural, 195
   population size and, 174–175
   symbolic behaviors and pace of,
   191–194
(IIT) integrated information theory, 224,
   225
intelligence
   ecological hypothesis of, 102–103
   environmental factors, 206, 214
   genetic influences, 207, 214
   intelligent animals, 52, 185
   mammalian lineages, 52
   selection for, 184
invertebrates, 41, 234n
ion channels, 30, 35, 36, 234n
iPSCs (induced pluripotent stem cells),
   220, 221
iridium, 56
Iwarrika (mythological monkey figure),
   113
Izard, Carroll, 147

Jacka, Felice, 182–184
Jarrett, Christian, 111
jaw, 86, 166
Jerison, Harry, 62
Jurassic period, 47, 51

Kahama killings, 116, 119
Kano, Fumihiro, 108
Kano, Takayoshi, 126–127, 130
Kasekela chimpanzee community, 116,
   117–118
Khoisan people, 82
Kibale National Park (Uganda), 117
King, Barbara J., 151, 164, 167, 172, 185
King, William, 84
Koch, Christof
   on artificial intelligence, 226
   on effects of modern technology, 217

on genetic engineering of brains, 221–222
on integrated information theory (IIT), 224, 225
on pace of cultural evolution, 215
on selection pressure of intelligence, 213
Kokolopori Bonobo Research Project (DRC), 130
Kolbert, Elizabeth, 94
Krupenye, Christopher, 108
Kübler-Ross, Elisabeth, 151

Laetoli site, 75–77
Lahn, Bruce, 206
Laland, Kevin, 191
Lamarck, Jean-Baptiste, 215
lamprey fish, 42, 43
lancelets, 41
land bridges, 66, 179, 201
language
    ape language research, 136
    chimpanzees, 142, 236
    chirping birds, 137, 156
    in emotional evolution, 147–148
    genetic influences, 137, 139
    hyoid bone, 137
    monkey anatomy, 139–140
    musicality in, 196
    neurocircuitry of, 141, 144
    proto-languages, 138
    as stimulus for brain evolution, 136
    symbolic, 99, 140, 144, 159
laughter, 142–143
leaf-eating species (foliovores), 102
Leakey, Louis, 187, 190
lemurs, 57, 59, 102
life, beginning of, 27–28, 30–31
lifespan, prolonged, 59, 61
Linnaeus, Carl, 80, 235n
lions, prey preferences, 167, 168, 237n
livestock, 201
Lola ya Bonobo (Democratic Republic of the Congo), 129
longevity and dietary practices, 181, 182

lorises, 57, 59, 102
love, neurobiology of, 148–150
Lovejoy, C. Owen, 157–158
Lucy, 74, 75, 187
lungs, 44
Luo, Zhe-Xi, 51–53

macaques
    creative behavior, 185
    encephalization quotient, 62
    frontal pyramidal branching, 92
    neural network, 109–110
    vocal tract architecture, 139–140
MacLean, Paul, 48–49, 50
Makgadikgadi Basin (Botswana), 83
males
    alpha, 127, 131, 198–199, 204, 213
    behavior adopted by female bonobos, 132
    genetic diversity, 212–213, 237–238n
    male and female sizes, 75, 86
    patrilineal groups, 203, 237–238n
mammals
    child-rearing, 51
    defined, 50–51
    diet, 53
    distinct branches, 51
    explosion in diversity, 57
    intelligent, 52
    mammalian lineages, 52
    neomammalian (new mammal) brain, 48, 49, 50
    nocturnal, 52, 56, 57, 59, 60
    placental, 51, 52
Man the Tool-Maker (Oakley), 189
Marean, Curtis, 173–174, 175, 178, 193
Marine Isotope Stage 6 (MIS6), 173, 174, 237n
marmoset monkeys, 92, 235n
marsupials, 51, 52
mass casualties, 212–213
Matama, Hilali, 115, 116
materialism, philosophy of, 224
matrilineal human groups, 202
McNernie, Janice, 19

meat consumption
    becoming carnivorous, 12, 166–167
    bone marrow, 168
    brain size and, 12, 166–167, 168–169
    cooked meat, 169
    expensive tissue hypothesis, 168–169
    fossil evidence of butchering,
        167–168
    organ meat, 167
    prey preferences, 167, 168, 237n
Mediterranean Diet, 182, 183–184
mental health
    dietary influences, 170, 176–177,
        181–184
    genetic influences, 177, 207–209
mesotherm, 58
Metzinger, Sam, 222–223
Mijares, Armand, 82
MIND diet, 184
Miocene epoch, 66–67
missing link, 41, 72, 234n
Mitani, John, 117–118, 128, 165
mitochondrial Eve hypothesis, 65, 82
Miyagawa, Shigeru, 137
"molecular clock," 64
mollusks, 34–35, 174
monkeys
    ability to access food, 164–165
    anatomy and language, 139–140
    F5 brain region, 140
    identification, 66
    as mythological figures, 113
monogamy, 204, 236n
monotremes, 51, 52
Montserrat, Jordi Paps, 39
morality, 122, 128, 133
mortality, 181, 213–214
mouth, 166
MRI scans, 3, 49, 80, 83, 109,
    110, 183
Muir, John, 30, 224
multicellular life, 31–32, 34
muscles
    for chewing, 170
    contraction of, 37, 234n

muscle in consumed meat, 169
    of speech, 37, 140
    tightening of, 42, 145–146
musical abilities, 205–206
music and rhythm, 194–196
myelin, 35, 36
myth, shared, 198–199

Native Americans, 201
natural selection
    cultural influences vs., 122
    defined, 21, 29
    especially influential factors, 87–88
    evolutionary byproducts, 226
    genetic mutations, 29, 138, 193
    group selection, 104–105
    rate of, 212
    reproduction and survival benefits,
        146, 158
nature vs. nurture, 137–138, 205–206
Neanderthal Museum (Mettmann,
    Germany), 85
Neanderthals
    brain size, 84, 85–86
    cave art, 85, 194, 235n
    extinction of, 94, 95
    hyoid bone, 137
    improvement of reputation, 85
    intelligence, 85
    language, 137
    in popular culture, 84–85
    Sapien / Neanderthal comparisons,
        85–86, 91, 137, 197
neomammalian (new mammal) brain,
    48, 49, 50
nervous system
    autonomic nervous system, 42
    central nervous system, 40, 47, 234n
    evolution of, 35–37, 40, 234n
    parasympathetic, 42, 146
    sympathetic, 42, 145–146
neural crest cells, 156–157
neurocircuitry
    axons, 35, 36, 62, 63
    of communication, 140, 144

defined, 62

dendrites, 35, 62

in the modern human brain, 92–94

process, 35–36

of reward system, 35, 148–149

of social primates, 109–110, 128

synapses, 35–36, 207

of toolmaking, 188–189

visual circuitry, 62

neuroimaging, 49, 63, 109–111, 141

neurons

benefits of omega fats, 176

in early animals, 39

mirror neurons, 110–111, 140

precursor cells, 234n

pyramidal, 92–93

spindle, 92, 110

neurotransmitters, 35–37, 121–122,
157–158, 234n

newt neuroanatomy, 46–47

New World monkeys, 58, 59

Ngamba Island Chimpanzee Sanctuary
(Uganda), 129

nocturnal mammals, 52, 56, 57, 59, 60

non-adaptive traits and abilities, 156, 196

norepinephrine, 122, 234n

Oakley, Kenneth Page, 189

obstetrical dilemma, 90–91, 148, 172

ochre, uses, 192-193, 194, 195, 197

octopus neuroanatomy, 34–35

Okinawan diet, 182

Oldowan tool industry, 187

Old World monkeys, 58, 59, 62, 66

olfactory bulbs, 52, 60

omega-3s, 176–178, 180–181

omnivorism, 11–12, 102, 184

Orrorin tugenensis, 72

Otavia antiqua, 33

ownership, 202

oxytocin, 111, 121, 149

Paleo diet, 183–184

Paleolithic Technology Laboratory
(Emory University), 188

paleomammalian (old mammal) brain,
48, 49, 145

panpsychism, 224

Panzee. *See* Chim and Panzee

Paranthropus, 76, 164, 236n

parenting, 105

paternity, ambiguous, 131–132, 204

patrilineal groups, 203, 237–238n

Pauling, Linus, 64

Permian extinction, 47, 55

Peterson, Dale, 118

Peugeot, 198

photosynthesis, 31, 56–57

Pinker, Steven, 119–120, 122, 137–138,
195

Pinnacle Point (South African coast),
173–174, 193–194

placental mammals, 51, 52

Pleistocene epoch, 81, 86, 91

plesiadapiforms, 57

Pobiner, Briana, 167–168

polyunsaturated fatty acids (PUFAS),
176, 177

population

diversity, 203

fluctuations, 173, 175, 202

male / female ratios, 212

mass casualties, 212–213

preadaptations, 106, 138, 142, 166

prefrontal cortex

Brodmann's Area 10, 93–94

location, 189

negotiating with the amygdala, 147

planning and abstract thought, 92, 188

sense of reward, 149

prey preferences, 167, 237n

primates

arboreal, 58–59

early, 57–58

intelligence in, 52

rodentesque, 57–58, 59

transitional species, 57

"wet-nosed" and "dry-nosed," 58, 59

Proconsul, 66

Project Chimps, 23

prosimians, 57
proto-languages, 138, 144
proto-mammals, 51
proto-synapses, 37
PUFAS (polyunsaturated fatty acids), 176, 177
purines, 27, 28
pyrimidines, 27, 28

qualia, 223
Quaternary period, 163

rafting, 58, 66, 179
Raghanti, Mary Ann, 157–158
Ramachandran, Vilayanur, 111
Ramón y Cajal, Santiago, 218
Ramsey, Drew, 11
rangeomorphs, 32, 33
red (color), 192–193
religion and spiritualism, 196–199
reptiles, 46, 47, 52
reptilian brain, 48, 49, 50, 235n
research, correlational, 180, 184
reward system, 35, 148–149
rhythm, 195
ritual, 197, 198
Rizzolatti, Giacomo, 110, 140
RNA, 27–28, 29, 219
Rosati, Alexandra, 169
Rousseau, Jean-Jacques, 115
Rukwapithecus fleaglei, 66
Ryan, Christopher, 119, 203, 217

Sagan, Carl, 113
Sahelanthropus tchadensis, 71
San Diego Zoo, 17, 23
Sapiens
    descendants, 82–83
    marine foraging, 174, 175
    migration failures, 82
    population fluctuations, 173
    Sapien / Neanderthal comparisons, 85–86, 91, 137, 197
Sapiens (Harari), 194, 199
Savage, Thomas Staughton, 114

Scandinavian diet, 182
scavenging, 167, 168
schizophrenia, 207, 208–209
seafood, 174, 175–177
self-domestication
    control of aggressive behavior, 152, 154, 213
    role of neurotransmitters, 157–158
    shrunken jaw as harbinger of, 166
self-help sensationalism, 216
The Selfish Gene (Dawkins), 104
Semendeferi, Katerina, 92–94, 141
serotonin, 35, 122, 157–158, 234n
sex, evolution of, 31
Sex at Dawn (Ryan and Jethá), 204
sex chromosomes, 237–238n
sexual behavior
    desire, 149
    polygamy, 204
    sex drive, 146
    strategic promiscuity, 131–132
sexual dimorphism, 74, 86
Seymour, Roger, 78
sharks, 42
Shubin, Neil, 44–45, 50, 53
silver foxes, 154–155
Sima de los Huesos site, 91
simians, 58, 60, 61
single-celled organisms, 30–31, 32–33
The Sixth Extinction (Kolbert), 94
skeleton, 41
smelling, 52, 56, 60
social behavior. *See also* communal living
    bonding, 143–144
    eusociality, 103–104
    grooming, 99–101
    helping, 105
    selection pressures, 103
    tribalism, 107
    virtual communication, 216–218
social brain hypothesis, 101–103, 141–142
social intelligence
    ape / human toddler social intelligence study, 108–109
    beginnings of, 100–101

birds, 236n
continuum in, 109
cooperation, 99–100, 107, 164, 199
inferring the intentions of others, 100,
   106, 107–108, 110, 111, 120, 236n
larger group size and, 141–142
neurotransmitters, 157
selection for, 105
social intelligence hypothesis, 102, 190,
   236n
social stratification, 203
spandrels, 226
speech, 141
spine, 41, 42
spiritualism and religion, 196–199
sponges, 25–27, 32–33
stem cells, 220–222
Stout, Dietrich, 188, 189–190
strepsirhines ("wet-nosed" primates),
   58, 59
stress, 145–146
stress hormone, 145, 155–156
striatum, 149, 157
sulci, 62
sulci (folds), 62, 63, 79, 188
Surbeck, Martin, 130–133
survival of the fittest, 29, 154
Survival of the Friendliest (Hare), 154
Sutherland, John, 27–28
symbolic cognition
   artistic expression, 85, 193–196, 235n
   language, 99, 140, 144, 159
   Neanderthals, 85, 95
   pace of innovation and, 186–187,
      191–194
   use of ochre, 192–194
synapsids, 47, 51, 52, 235n
synaptic pruning, 207, 208

tarsiers, 59
Tattersall, Ian
   on artificial intelligence, 227
   on danger for early hominins, 77
   on epigenetic effects of virtual
      communication, 218

on evolutionary innovation, 174–175
on innovation in small populations,
   174–175
on language, 138, 139, 144, 191–192
museum displays by, 73, 74, 75, 85
museum exhibits curated by, 73, 85
on neurocircuitry, 92
on recent human history, 218
on selection, 87–88
on use of fire, 171
Taung Child, 72, 73, 75, 78–79, 90, 167
teeth, 53, 86, 158, 166
tetrapods, 46
thalamus, 42
theism, 199
theory of mind (TOM), 107–111, 236n
therapsid lineage, 46, 47, 51
throwing, 86–87
Tiktaalik, 44–45
TOM (theory of mind), 107–111, 236n
Tomasello, Michael, 105, 107, 108–109,
   202, 236n
toolmaking, 187–192
Toumaï, 71–72
trees. *See* arboreal life
trilobites, 40, 55
tripartite brain, 40
triune brain model, 48–49, 50
Troglodytes gorilla, 114
tropical habitats, 77, 165
Tucker-Drob, Elliot, 205
Turing, Alan, 224
Turing Test, 224
twin gene-edited babies, 219–220
twin studies, 205, 214
2001: A Space Odyssey (movie), 224

United States Preventive Services Task
   Force, 181
upright posture, 71–73, 75–77, 114

VENs (von Economo neurons), 92
Venus figurines, 194
vertebrates
   ancestors of, 34

brain development, 40, 42, 50
establishment on land, 46
evolution as predators, 43
missing link between invertebrates
    and vertebrates, 41, 234n
newt anatomy, 46
Vinther, Jakob, 43
violence
    in chimpanzees, 20, 115–119, 121,
        122–123, 152
    control of, 122, 152, 213
    decline of, 119–120
    early-life abuse, 121, 236n
    genetic influences, 121–122
    killing behavior, 20, 115–119
    origins of, 118–119
    philosophers' views, 115
    rates of, 120–121
    reactive and proactive, 152
    socioeconomic factors, 121
    toward females / women, 122–123
vision
    color vision, 60
    improvement through selection
        pressure, 52, 59–60
    night vision, 52, 56, 60
    retinal cones, 60, 192–193
    trichromatic, 193
voice, human, 194
volcanic eruptions, 55
von Economo neurons (VENS), 92

Wallace, Alfred, 13, 29, 233n
war, 118–119
Warneken, Felix, 169
Wells, H.G., 85
Wernicke's area, 140, 141, 144
Western diets, 183
whales, 151
Wheeler, Peter, 168–169
whiskers, 56
Why Only Us (Chomsky), 137
Williams-Beuren syndrome, 157
Wilson, Allan, 64–65, 71, 101

Wilson, E.O.
    on community in evolutionary
        process, 105–106, 108, 143
    on the consequences of agriculture,
        204
    on eusociality, 103–104
    on interaction of selection and
        culture, 228
    on terrestrial living, 70
Wonderwerk Cave (South Africa), 169
Wrangham, Richard
    on aggression in chimps, 118–119
    on calories, 179
    Catching Fire: How Cooking Made Us
        Human, 169, 172
    on chimpanzee vs. bonobo behavior,
        129–130, 156
    on cultural innovation, 195
    on domestication process, 156
    grooming and playing behaviors
        study, 129
    on human social networks, 102
    on nice aspects of terrible people, 133
    on the raw food movement, 170
    on role of cooking, 172
    on self-domestication, 152, 154, 213
    on use of fire, 169, 171, 172
writing system, 197
Wyman, Jeffries, 114

xenophilia, 20
xenophobia, 106–107

Yamanaka, Shinya, 220
Yerkes, Robert, 125–126
Yerkes National Primate Research Center
    (Atlanta), 127
Your Inner Fish (Shubin), 44, 53
Yucatán Peninsula, 56

zombie problem, 223–224
Zuckerkandl, Émile, 64